CHEMISTRY
TEACHERS' GUIDE II
TOPICS 12 TO 18

Revised Nuffield Advanced Science

General editor
Revised Nuffield
Advanced Chemistry
B. J. Stokes

Associate editor,
inorganic chemistry
A. J. Furse

Associate editor,
organic chemistry
M. D. W. Vokins

Associate editors,
physical chemistry
**D. H. Mansfield,
Professor
E. H. Coulson,
Jon Ogborn**

Editor of this book
B. J. Stokes

Contributors to this book
A. W. B. Aylmer-Kelly
Professor E. H. Coulson
B. E. Dawson
A. J. Furse
John Holman
A. J. Malpas
D. H. Mansfield
Professor D. J. Millen
Jon Ogborn
J. G. Raitt
B. J. Stokes
M. D. W. Vokins

Consultant on safety matters
Dr T. P. Borrows

CHEMISTRY TEACHERS' GUIDE II

Topics 12 to 18

Revised Nuffield Advanced Science
Published for the Nuffield–Chelsea Curriculum Trust
by Longman Group Limited

Longman Group Limited
Longman House, Burnt Mill, Harlow, Essex CM20 2JE, England
and Associated Companies throughout the World

First published 1970
Revised edition first published 1984
Copyright © 1970, 1984, The Nuffield–Chelsea Curriculum Trust

Design and art direction by Ivan Dodd
New diagrams by Oxford Illustrators Limited

Filmset in Times Roman and Univers
Printed in Great Britain
by Richard Clay (The Chaucer Press) Bungay

ISBN 0 582 35363 7

All rights reserved, no part of this publication may be reproduced, stored in a retrieval system, or transmitted in any form or by any means – electronic, mechanical, photocopying, or otherwise – without the prior written permission of the Publishers.

Note
All references to the **Book of data** are to the **revised** edition which is part of the present series.

Contents

Foreword   page *vi*

Introduction *viii*

Topic 12   Equilibria: gaseous and ionic *1*

Topic 13   Carbon compounds with acidic and basic properties *42*

Topic 14   Reaction rates – an introduction to chemical kinetics *63*

Topic 15   Redox equilibria and free energy *92*

Topic 16   The Periodic Table 4: the transition elements *137*

Topic 17   Synthesis: drugs, dyes, and polymers *154*

Topic 18   The Periodic Table 5: the elements of Groups III, IV, V, and VI *177*

Appendix 1   Nomenclature, units, and abbreviations *192*

Appendix 2   Apparatus *194*

Appendix 3   Models and other visual aids *212*

Appendix 4   Possible alternative pathways through the course *224*

Reference sources and bibliography *226*

Schools which took part in the trials *230*

Index *231*

Foreword

When the Nuffield Advanced Science series first appeared on the market in 1970, they were rapidly accepted as a notable contribution to the choices for the sixth form science curriculum. Devised by experienced teachers working in consultation with the universities and examination boards, and subjected to extensive trials in schools before publication, they introduced a new element of intellectual excitement into the work of A-level students. Though the period since publication has seen many debates on the sixth form curriculum, it is now clear that the Advanced Level framework of education will be with us for some years in its established form. Although various proposals for change in structure have not been accepted, the debate to which we contributed encouraged us to start looking at the scope and aims of our A-level courses and at the ways they were being used in schools. Much of value was learned during those investigations and has been extremely useful in the planning of the present revision.

The revision of the chemistry series under the general editorship of B. J. Stokes has been conducted with the help of a committee under the chairmanship of Malcolm Frazer, Professor of Chemical Education, University of East Anglia. We are grateful to him and to the committee. We also owe a considerable debt to the London Examinations Board which for many years has been responsible for the special Nuffield examinations in chemistry and to the subject officer, Peter Thompson, who has been an invaluable adviser on these matters.

The Nuffield–Chelsea Curriculum Trust is also grateful for the advice and recommendations received from its Advisory Committee, a body containing representatives from the teaching profession, the Association for Science Education, Her Majesty's Inspectorate, universities, and local authority advisers; the committee is under the chairmanship of Professor P. J. Black, academic adviser to the Trust.

Our appreciation also goes to the editors and authors of the first edition of Nuffield Advanced Chemistry, whose work, under the direction of E. H. Coulson, the project organizer, made this one of our most successful and influential ventures into curriculum development. Ernest Coulson's team of editors and writers included A. W. B. Aylmer-Kelly, Dr E. Glynn, H. R. Jones, A. J. Malpas, Dr A. L. Mansell, J. C. Mathews, Dr G. Van Praagh, J. G. Raitt, B. J. Stokes, R. Tremlett, and M. D. W. Vokins. A great part of their original work has been preserved in the new edition, on which several of them have acted as consultants.

I particularly wish to record our gratitude to Bryan Stokes, the General Editor of the revision. As a member of the original team he has an unrivalled understanding of the aims and scope of the first edition and as a practising teacher he possesses a particular awareness of the needs of pupils and teachers which has enriched the work of the revision. To him, to the editors working with him, A. J. Furse (Inorganic Chemistry), M. D. W. Vokins (Organic Chemistry), J. A. Hunt who is responsible for the Special Studies, and to the team responsible for the Physical Chemistry sections, Professor P. J. Black, J. S. Holman, D. H. Mansfield, Professor D. J. Millen, and Jon Ogborn, we offer our most sincere thanks.

I would also like to acknowledge the work of William Anderson, publications manager to the Trust, his colleagues, and our publishers, the Longman Group, for their assistance in the publication of these books. The editorial and publishing skills they contribute are essential to effective curriculum development.

K. W. Keohane,
Chairman, Nuffield–Chelsea Curriculum Trust

Introduction

This book is volume II of the two-volume *Teachers' guide* to Revised Nuffield Advanced Chemistry. It covers Topics 12 to 18, the seven topics of *Students' book II*. These topics, together with a Special Study, constitute the work suggested for the second year of the course.

In this revised course, the basic aims and ideas of the first edition, the outline of the topics, and much of the detail remain. Since the first edition was published, however, many developments in chemistry have taken place. New discoveries have been made; industrial processes have advanced; there have been changes in the way many chemicals are named, in the units in which many quantities are measured, in safety practices, and in a number of other matters. This second edition takes account of all these.

For a full discussion of the nature of the revision and a description of the content of the revised course, the reader is referred to the Introduction to *Teachers' guide I*.

Contributors

Many people have contributed to this book. Final decisions on the content and method of treatment used in the first edition were made by the Headquarters team, who were also responsible for assembling and writing the material for the several draft versions that were used in school trials. The Headquarters team consisted of E. H. Coulson (organizer), A. W. B. Aylmer-Kelly, Dr E. Glynn, H. R. Jones, A. J. Malpas, Dr A. L. Mansell, J. C. Mathews, Dr G. Van Praagh, J. G. Raitt, B. J. Stokes, R. Tremlett, and M. D. W. Vokins.

The revision has been undertaken largely by three working groups, whose members were:

Inorganic chemistry: A. J. Furse (chairman), K. W. Badman, M. C. V. Cane, C. Nicholls, and D. Russell.

Organic chemistry: M. D. W. Vokins (chairman), J. J. Eggleton, G. H. James, and Professor D. J. Waddington.

Physical chemistry: Professor M. J. Frazer (chairman), Professor P. J. Black, Dr T. P. Borrows, John S. Holman, D. H. Mansfield, Professor D. J. Millen, and Jon Ogborn.

Advice on safety matters has been given by Dr T. P. Borrows, Chairman of the Safety Committee of the Association for Science Education.

This book has benefited greatly from the valuable help and advice that have been generously given by the teachers in schools and in universities and other institutions of higher education. In particular, the comments and suggestions of teachers taking part in the school trials, both of the original course, and of the revised topics, have made a vital contribution to the final form of the published material.

Finally, as editor, I should like to record my thanks to the Publications Department of the Nuffield–Chelsea Curriculum Trust for their help, and particularly to Mary de Zouche for her meticulous and painstaking attention to detail in the preparation of the manuscripts for publication, and to her colleagues, Deborah Williams, Hendrina Ellis, Nina Konrad, and Sarah Codrington.

B. J. Stokes

TOPIC 12
Equilibria: gaseous and ionic

OBJECTIVES

1 To consider the properties which characterize the equilibrium state.
2 To study the effects of conditions on the proportions of the reactants and products.
3 To consider entropy changes in an equilibrium reaction.

CONTENT

12.1 General introduction. The Equilibrium Law. Examples of equilibria, characteristics of the equilibrium state; relative concentrations under equilibrium conditions; experimental investigations of some equilibria; the equilibrium law and the equilibrium constant.

12.2 The effect of pressure and temperature on equilibrium. The equilibrium constant expressed in terms of partial pressures; relationship between K_c and K_p; the effect of pressure on equilibrium; the effect of a change of temperature on the equilibrium constant; Le Châtelier's principle applied to changes of concentration, pressure, and temperature.

12.3 Heterogeneous equilibria. Solutions and solubility product; other heterogeneous equilibria.

12.4 Acid–base equilibria. Acids as proton donors and bases as proton acceptors; the ionization constant for water; acid–base systems and their chemistry; the strengths of acids and bases; pH values; the use of pH meters; change of pH during acid–base titrations examined experimentally.

12.5 Buffer solutions and indicators. Buffer solutions; calculation of pH; acid–alkali indicators; experimental measurement of K_a for an indicator and a weak acid; pK_a values. Background reading: 'Acid–base chemistry in the human body'.

12.6 Equilibrium and zero total entropy change. Entropy balance sheets for reversible systems; equilibrium constant and free energy.

TIMING

About two and a half weeks.

Topic 12 Equilibria: gaseous and ionic

INTRODUCTION

This is the first Topic in this course to deal with the principles governing equilibrium reactions, although examples of such reactions have been met on a number of previous occasions. In this Topic, the Equilibrium Law is introduced as the result of experiment, and its implications are considered for a number of gaseous, liquid, and ionic equilibria, including acid-base reactions.

In the first edition of this course, acid-base reactions were introduced after redox equilibria in Topic 15. This enabled a comparison to be made between the electron transfer of redox systems and the proton transfer of acid-base systems. Furthermore, if acid-base reactions are studied after redox equilibria, it is possible to explain the working of the pH meter when the instrument is used, for the students will understand the way in which the e.m.f.s of cells vary with the concentration of ions, and will appreciate how the e.m.f. of a cell can be used to measure the concentration of hydrogen ions.

With the reorganization that has taken place in this second edition, the study of acid-base equilibria has been brought forward to this Topic, where it can be compared with other equilibrium studies. This has been done so that when Topic 15 is reached, attention can be concentrated on the final development of the ideas of entropy and free energy in redox systems (the revised treatment of the subject matter of Topic 17 of the first edition).

It is, of course, open to the teacher to choose whichever order of Topic that is preferred.

12.1
GENERAL INTRODUCTION. THE EQUILIBRIUM LAW

Objectives

1 To discuss the properties which characterize the equilibrium state.
2 To develop the Equilibrium Law as an empirical relationship arising from experimental studies.

Timing

About two periods.

Suggested treatment

For this treatment overhead projection transparencies numbers 104–107 will be useful.

In Topic A19 of Revised Nuffield Chemistry, the following equilibria are studied:

12.1 General introduction. The Equilibrium Law

1 $Cl_2(g) + ICl(l) \rightleftharpoons ICl_3(s)$
2 $Br_2(aq) + H_2O(l) \rightleftharpoons H^+(aq) + Br^-(aq) + HOBr(aq)$
3 $2CrO_4^{2-}(aq) + 2H^+(aq) \rightleftharpoons Cr_2O_7^{2-}(aq) + H_2O(l)$
4 $BiCl_3(aq) + H_2O(l) \rightleftharpoons BiOCl(s) + 2H^+(aq) + 2Cl^-(aq)$
5 $Ag^+(aq) + Fe^{2+}(aq) \rightleftharpoons Ag(s) + Fe^{3+}(aq)$

Students should be reminded of these experiments and, if necessary, a selection of them could be demonstrated; **1, 3,** and **5** are probably the most informative at this stage. This will provide some further examples of reversible changes, and of changes which do not go to completion, that is, reactions in which some of the original materials are still present when the system has reached a stable state.

Discussion can now be concentrated on the following points:

1 A stable state of equilibrium can be set up only in a *closed* system – one which contains a constant amount of matter but can exchange energy with its surroundings (the contents of a stoppered vessel), or in an *isolated* system – one which contains a constant amount of both matter and energy (for example the contents of a sealed and insulated vessel). Stable equilibrium cannot be attained in an open system, that is, one that can exchange both matter and energy with its surroundings.

2 If allowed to stand long enough all of these systems will reach a position of equilibrium whether the starting materials are the substances on the left of the $\rightleftharpoons$ or on its right. The symbol '$\rightleftharpoons$' indicates that equilibrium can be approached from either direction. Changes of this kind are called *reversible reactions*. Equilibrium may be attained very rapidly, or very slowly. The possibility that a system may appear to be in the equilibrium state when it is, in fact, changing at a very slow rate, makes the criterion that equilibrium is approachable from either direction important.

3 The equilibrium state is characterized by constancy of *intensive properties* in a closed or isolated system. (We shall be concerned only with closed, not with isolated systems.) Intensive properties are those which are independent of the total amount of matter in the system, such as depth of colour per unit thickness of layer, density, pressure, and concentration. Properties such as mass, volume, and internal energy, which vary with the total amount of matter, are called *extensive properties*.

4 The approach to equilibrium from either direction suggests that the equilibrium state is dynamic rather than static. Evidence for this is normally obtainable only by use of radioisotopes as tracer elements. An experiment using a radioisotope for showing the dynamic equilibrium between a solute and its saturated solution is given in Revised Nuffield Chemistry, *Teachers' guide II*, pages 246–247, as experiment A19.3.

Relative concentrations under conditions of equilibrium

1 *The distribution of a solute between two immiscible solvents.*

Instructions for investigating the distribution of ammonia between 1,1,1-trichloroethane and water are given as Experiment 12.1; students can use the results of their experiment to obtain the Equilibrium Law. Experimental investigations in this area of the subject are time-consuming, and for further examples it is recommended that data obtained by others are used.

Tables showing equilibrium concentrations for the reaction between hydrogen and iodine in the gas phase, and the hydrolysis of ethyl ethanoate, are given in the *Students' book*. These should be discussed on the lines indicated by the questions and comments following each table. The calculations based on the figures in tables 12.1, 12.2 and 12.3 of the *Students' book* could be shared amongst the students, so that results for a whole table can be obtained in a short time. For the information of the teacher complete versions of the tables are given below, together with notes on the questions and comments for all these systems.

EXPERIMENT 12.1
The distribution of ammonia between water and 1,1,1-trichloroethane

Each student or pair of students will need:
Glass-stoppered bottle, 250 cm^3
Measuring cylinder, 100 cm^3
2 pipettes, 25 cm^3
Pipette filler
2 conical flasks, 250 cm^3
Separating funnel, 100 cm^3
2 burettes, 50 cm^3
2 burette stands and clamps
6M ammonia solution, 75 cm^3
1,1,1-trichloroethane, 75 cm^3
Methyl orange indicator
4.0M hydrochloric acid, 150 cm^3
0.10M hydrochloric acid, 150 cm^3

A safety pipette filler MUST be used in this experiment. Students must not be allowed to mouth pipette any of these solutions.

It is essential that the stopper of the glass-stoppered bottle is a good fit. This should be tested using water *before* the organic solvent is added.

Although 1,1,1-trichloroethane is less toxic than trichloromethane, the solvent traditionally used for this experiment, it must still be treated with care; a well-ventilated laboratory is desirable. Also, thought needs to be given to the disposal of the relatively large quantities of waste solvent which will accumulate.

The best method of disposal is probably slow evaporation in the open air in an area inaccessible to students. Large quantities of 1,1,1-trichloroethane should not be poured down the sinks; first this may not be allowed by the Water Authority and second, there is a risk of the liquid accumulating in the traps in the laboratory drainage system. The disposal of waste liquids is discussed in Chapter 13 of *Topics in Safety*, Association for Science Education, 1982.

Procedure

Full details of the experimental procedure are given in the *Students' book*. Two questions are asked at the end of the instructions. The answers to be expected are:

1 No units are given for the figure in the last column of the table because the units cancel when the ratio is calculated.

2 The ratio of concentrations in the two layers is nearly constant, but tends to increase slightly as the actual concentrations decrease. Teachers will appreciate that this is due to the equilibrium

$$NH_3(aq) + H_2O(l) \rightleftharpoons NH_4^+(aq) + OH^-(aq)$$

which is set up in the water layer. It reduces the *effective* NH_3 concentration in the water layer which means that the value of $[NH_3(H_2O)]_{eqm}$ determined is too high and thus the concentration ratio is high also. The effect should be more marked at lower ammonia concentrations. It may however not be necessary (or desirable) to raise this additional complication when discussing this equilibrium with the students.

Specimen results for this experiment are given on overhead projection transparency number 105.

2 *The hydrogen iodide equilibrium*

Completed versions of the tables given in the *Students' book*, using the relationships suggested in the *Students' book*, are as follows.

$[H_2(g)]_{eqm}$ /mol dm^{-3}	$[I_2(g)]_{eqm}$ /mol dm^{-3}	$[HI(g)]_{eqm}$ /mol dm^{-3}	$\dfrac{[HI(g)]_{eqm}}{[H_2(g)]_{eqm}[I_2(g)]_{eqm}}$ /dm^3 mol^{-1}	$\dfrac{[HI(g)]_{eqm}^2}{[H_2(g)]_{eqm}[I_2(g)]_{eqm}}$
4.56×10^{-3}	0.74×10^{-3}	13.54×10^{-3}	4.01×10^3	54.3
3.56×10^{-3}	1.25×10^{-3}	15.59×10^{-3}	3.50×10^3	54.6
2.25×10^{-3}	2.34×10^{-3}	16.85×10^{-3}	3.20×10^3	53.9

Table 12.1
Results obtained by heating hydrogen and iodine in sealed vessels.

$[H_2(g)]_{eqm}$ /mol dm^{-3}	$[I_2(g)]_{eqm}$ /mol dm^{-3}	$[HI(g)]_{eqm}$ /mol dm^{-3}	$\dfrac{[HI(g)]_{eqm}}{[H_2(g)]_{eqm}[I_2(g)]_{eqm}}$ /dm^3 mol^{-1}	$\dfrac{[HI(g)]^2_{eqm}}{[H_2(g)]_{eqm}[I_2(g)]_{eqm}}$
0.48×10^{-3}	0.48×10^{-3}	3.53×10^{-3}	15.3×10^{-3}	54.1
0.50×10^{-3}	0.50×10^{-3}	3.66×10^{-3}	14.6×10^{-3}	53.5
1.14×10^{-3}	1.14×10^{-3}	8.41×10^{-3}	6.47×10^{-3}	54.4

Table 12.2
Results obtained by heating hydrogen iodide in sealed vessels.

The expression $\dfrac{[HI(g)]^2_{eqm}}{[H_2(g)]_{eqm}[I_2(g)]_{eqm}}$ gives the better constant in all the cases quoted above. It should now be related to the stoicheiometric equation for the reaction. By convention, the concentrations of products are put into the numerator and those of the reactants into the denominator.

The symbol K_c can now be introduced (*the equilibrium constant* when calculated from concentration terms).

Notes on the tables

a Values of K_c can be based on the equation

$$\tfrac{1}{2}H_2(g) + \tfrac{1}{2}I_2(g) \rightleftharpoons HI(g)$$

that is $K_c = \dfrac{[HI(g)]_{eqm}}{[H_2(g)]^{\frac{1}{2}}_{eqm}[I_2(g)]^{\frac{1}{2}}_{eqm}}$

The values will be the square roots of those given in tables 12.1 and 12.2, that is about 7.3. This means that *whenever a value for an equilibrium constant is quoted the equation on which it is based must be indicated*.

b The average value of K_c correct to 2 significant figures for the results in tables 12.1 and 12.2 is 54.2. If an equilibrium mixture of hydrogen, iodine, and hydrogen iodide at 698 K contains 0.5 mole of hydrogen and 5.42 moles of hydrogen iodide, the amount of iodine present can be determined from the above value of the equilibrium constant. If the total volume of the mixture is V cubic decimetres, the equilibrium expression is

$$\dfrac{\left(\dfrac{5.42}{V}\right)^2}{\left(\dfrac{0.5}{V}\right)\left(\dfrac{[I_2(g)]_{eqm}}{V}\right)} = 54.2$$

The volume of the system is not needed to obtain an answer as the expression reduces to

$$\frac{5.42^2}{0.5 \times [I_2(g)]_{eqm}} = 54.2$$

From which $[I_2(g)]_{eqm} = 1.08$ mole.

c The concentration of hydrogen and iodine is the same in each of the first two columns of table 12.2 because these results were obtained by heating hydrogen iodide in sealed vessels, and for every mole of hydrogen that is produced from hydrogen iodide, one mole of iodine is also produced.

3 *The equilibrium between ethyl ethanoate, water, ethanoic acid, and ethanol*

$$CH_3CO_2C_2H_5(l) + H_2O(l) \rightleftharpoons CH_3CO_2H(l) + C_2H_5OH(l)$$

The expression for the equilibrium constant is

$$K_c = \frac{[CH_3CO_2H(l)]_{eqm}[C_2H_5OH(l)]_{eqm}}{[CH_3CO_2C_2H_5(l)]_{eqm}[H_2O(l)]_{eqm}}$$

The complete version of table 12.3 including K_c calculated in this way, is given below. It shows that the experimental results confirm this choice of expression for K_c as the correct one.

Amount of ethyl ethanoate at equilibrium/mole	Amount of water at equilibrium/mole	Amount of ethanoic acid at equilibrium/mole	Amount of ethanol at equilibrium/mole	K_c
0.231	0.079	0.065	0.065	0.233
0.204	0.118	0.082	0.082	0.279
0.150	0.261	0.105	0.105	0.281
0.090	0.531	0.114	0.114	0.272

Table 12.3

There is more variation in the values of K_c in this example. It is not easy to obtain reactants for this system entirely free from water, hence the total amount of water present must be uncertain.

Figures in the table are in fact *amounts* of each substance present at equilibrium, and not their concentrations as given in tables 12.1 and 12.2. The volume term cancels (as with the previous example) and so it is not required for calculating K_c.

From a starting mixture of 3 moles of ethanol and 2 moles of ethanoic acid, the amount of ethyl ethanoate produced is obtained by substituting numerical values in the expression

$$K_c = 0.27 = \frac{[CH_3CO_2H(l)]_{eqm}[C_2H_5OH(l)]_{eqm}}{[CH_3CO_2C_2H_5(l)]_{eqm}[H_2O(l)]_{eqm}}$$

If x moles of ethanol react, then x moles of ethanoic acid will also react, and x moles of water and x moles of ethyl ethanoate will be formed. Thus at equilibrium

$$[CH_3CO_2H(l)]_{eqm} = \frac{2-x}{V} \text{ mol dm}^{-3}$$

$$[C_2H_5OH(l)]_{eqm} = \frac{3-x}{V} \text{ mol dm}^{-3}$$

$$[CH_3CO_2C_2H_5(l)]_{eqm} = \frac{x}{V} \text{ mol dm}^{-3}$$

$$[H_2O(l)]_{eqm} = \frac{x}{V} \text{ mol dm}^{-3}$$

and thus

$$0.27 = \frac{\left(\frac{2-x}{V}\right)\left(\frac{3-x}{V}\right)}{\left(\frac{x}{V}\right)\left(\frac{x}{V}\right)}$$

from which

$$0.73x^2 - 5x + 6 = 0$$

and $x = \dfrac{-(-5) \pm \sqrt{5^2 - (4 \times 0.73 \times 6)}}{2 \times 0.73} = 1.55$ (or 5.30) moles

This is

$1.55 \times 88 = 136.6$ g of ethyl ethanoate

Discussion of the tables of results of quantitative investigation of systems in equilibrium should lead to the generalization that for a change represented by

$$mA + nB \rightleftharpoons pC + qD$$

$$\frac{[C]^p_{eqm}[D]^q_{eqm}}{[A]^m_{eqm}[B]^n_{eqm}} = K_c \text{ (the equilibrium constant)}.$$

It should be noted that for each of the systems considered, the results were obtained at a fixed temperature.

Suggestion for homework

Answering questions 1–7 at the end of this Topic in the *Students' book*.

Summary

At the end of this section, students should be familiar with the following ideas concerning equilibria.

1 All systems tend to change in such a way that a state of equilibrium is attained, but the rate at which such changes take place varies considerably. A catalyst can be used to decrease the time needed for the attainment of equilibrium.

2 The state of equilibrium can be recognized by the constancy of intensive properties of the substances present.

3 A state of equilibrium can exist only in a closed or isolated system.

4 A state of equilibrium can be approached from either direction.

5 The state of equilibrium is dynamic in that processes on the molecular scale are taking place continuously but in such a way that they are in balance. Hence no change in bulk properties results.

6 The concentrations of reactants and products present at equilibrium are governed by the Equilibrium Law: 'at a given temperature, for a reaction represented by

$$mA + nB \rightleftharpoons pC + qD$$

the expression $\frac{[C]^p_{eqm}[D]^q_{eqm}}{[A]^m_{eqm}[B]^n_{eqm}}$ is constant for a particular system'. The symbol used for this constant is K_c (the equilibrium constant). The symbol $[A]_{eqm}$ represents the concentration of the species A in moles per cubic decimetre under equilibrium conditions. When the value of an equilibrium constant is quoted, the equation to which it refers should be indicated. Unless the number of particles is the same on both sides of the equation for the reaction, K_c must be given appropriate units.

7 In principle, all systems can be considered as reversible, and hence tend to move towards an equilibrium state.

12.2
THE EFFECT OF PRESSURE AND TEMPERATURE ON EQUILIBRIUM

Objectives

1 To introduce K_p, the equilibrium constant expressed in terms of partial pressures.

2 To investigate the effect of changes of pressure and of temperature on gaseous equilibria.

3 To introduce Le Châtelier's principle.

Timing

About two periods.

Suggested treatment

The equilibrium constant expressed in terms of partial pressures

When dealing with reactions involving gases, it is often found more convenient to use an equilibrium constant expressed in terms of pressure (K_p) rather than in terms of concentrations (K_c).

The use of partial pressures instead of concentrations for gas reactions will need to be explained to the students. This will involve quoting the law of partial pressures and showing that, from the ideal gas equation

$$pV = nLkT$$

$$\frac{n}{V} = [\text{gas}]$$

so that $[\text{gas}] = \dfrac{p}{LkT}$

Thus, for a given temperature, $[\text{gas}] \propto p$.

12.2 The effect of pressure and temperature on equilibrium

Note. Teachers will recall that the ideal gas equation was introduced in Topic 3 as

$$pV = nRT$$

with R being named as the gas constant. R was then shown to be the product of the Avogadro constant L and the Boltzmann constant k. From that point onward, the *Students' book* uses the ideal gas equation in the form

$$pV = nLkT$$

The *Students' book* provides examples of the calculation and use of K_p, and gives the relationship between K_p and K_c.

For the teacher's information, the way in which the relationship between K_p and K_c can be derived is set out below.

For the equilibrium

$$N_2O_4(g) \rightleftharpoons 2NO_2(g)$$

$$K_c = \frac{[NO_2(g)]^2_{eqm}}{[N_2O_4(g)]_{eqm}}$$

and $\quad K_p = \dfrac{p^2_{NO_2 eqm}}{p_{N_2O_4 eqm}}$

but $\quad [NO_2(g)]_{eqm} = \dfrac{p_{NO_2 eqm}}{LkT}$

and $\quad [N_2O_4(g)]_{eqm} = \dfrac{p_{N_2O_4 eqm}}{LkT}$

$$K_c = \frac{\left(\dfrac{p_{NO_2 eqm}}{LkT}\right)^2}{\left(\dfrac{p_{N_2O_4 eqm}}{LkT}\right)} = \frac{p^2_{NO_2 eqm}}{p_{N_2O_4 eqm}} \times \frac{1}{LkT}$$

$$= \frac{K_p}{LkT}$$

$$K_c LkT = K_p$$

The general expression is

$$K_p = K_c(LkT)^n$$

where $n =$ the number of gas molecules on the righthand side of the equation minus the number of gas molecules on the lefthand side of the equation. Where these are equal, as in

$$H_2(g) + I_2(g) \rightleftharpoons 2HI(g)$$

then $K_p = K_c$.

The effect of pressure on equilibrium

The expression for K_p can be used to find out what effect a change of pressure will have on a gaseous reaction equilibrium. This is discussed in the *Students' book*, resulting in the general statement that *increasing the pressure on a system in equilibrium produces a change which tends to decrease the volume*. Conversely, reducing the pressure favours the change which results in an increase in volume.

Thus for the sulphur trioxide equilibrium

$$2SO_2(g) + O_2(g) \underset{\text{pressure decrease}}{\overset{\text{pressure increase}}{\rightleftharpoons}} 2SO_3(g)$$

The *Students' book* contains some questions about the effect of pressure on certain equilibria. The expected answers are as follows.

1 The effect of an increase of pressure on the equilibrium

$$N_2O_4(g) \rightleftharpoons 2NO_2(g)$$

would be to move it to the lefthand side.

The effect of an increase of pressure on the equilibrium

$$CO(g) + 2H_2(g) \rightleftharpoons CH_3OH(g)$$

would be to move it to the righthand side.

The equilibrium

$$H_2(g) + Br_2(g) \rightleftharpoons 2HBr(g)$$

12.2 The effect of pressure and temperature on equilibrium

would not be affected. Neither would the equilibrium

$$CO_2(g) + NO(g) \rightleftharpoons CO(g) + NO_2(g)$$

2 In order to increase the yield of ethene from the equilibrium

$$C_2H_6(g) \rightleftharpoons C_2H_4(g) + H_2(g)$$

the pressure would have to be decreased.

The effect of temperature change on the value of the equilibrium constant for a reaction

The following table is given in the *Students' book*. The units of enthalpy change given in the table are kJ mol^{-1}. It may be necessary to explain to the students that mol^{-1} in this context refers to the amount of each substance as given in the equation.

$N_2(g) + 3H_2(g) \rightleftharpoons 2NH_3(g)$
$\Delta H^\ominus_{298} = -92 \text{ kJ mol}^{-1}$

T/K	K_p/atm^{-2}
500	3.55×10^{-2}
700	7.76×10^{-5}
1100	5×10^{-8}

$N_2(g) + O_2(g) \rightleftharpoons 2NO(g)$
$\Delta H^\ominus_{298} = +180 \text{ kJ mol}^{-1}$

T/K	K_p
700	5×10^{-13}
1100	4×10^{-8}
1500	1×10^{-5}

$2SO_2(g) + O_2(g) \rightleftharpoons 2SO_3(g)$
$\Delta H^\ominus_{298} = -197 \text{ kJ mol}^{-1}$

T/K	K_p/atm^{-1}
500	2.5×10^{10}
700	3×10^{4}
1100	1.3×10^{-1}

$H_2(g) + I_2(g) \rightleftharpoons 2HI(g)$
$\Delta H^\ominus_{298} = -9.6 \text{ kJ mol}^{-1}$

T/K	K_p
500	160
700	54
1100	25

Table 12.4
Variation of K_p values with temperature, and values of $\Delta H^\ominus_{298}$, for a series of reactions.

It will be seen that for an exothermic reaction K_p decreases as the temperature increases and for endothermic reactions the reverse is true. The *principle of Le Châtelier* can now be introduced as a general statement, which predicts qualitatively what effect a change in conditions will have on an equilibrium. This applies both to changes of concentration or pressure in which the system is adjusting itself so that the concentrations or partial pressures again satisfy the expressions for K_c or K_p; or to changes of temperature in which the system is adjusting itself *to a new value of K_c or K_p*.

The *Students' book* discusses how Le Châtelier's principle can be applied to changes of concentration, pressure, and temperature.

Suggestion for homework

Answering questions 8–11 at the end of this Topic in the *Students' book*.

Summary

From this section students should gain familiarity with the following:

1 How partial pressures can be used to calculate the equilibrium constant, K_p.

2 The effect of a change in pressure on a gaseous equilibrium can be determined from the expression for K_p.

3 How values of $\Delta H^\ominus_{298}$ for a reaction can be used to predict the effect of a change in temperature on an equilibrium.

4 The use of the principle of Le Châtelier in predicting qualitatively the effects of change of conditions on a system in equilibrium.

5 The very extensive range of values for the equilibrium constants for different reactions. A high value for K means that the reaction can proceed nearly to completion, a low value for K indicates that reactants will predominate when equilibrium is attained. It should be stressed, however, that the value of K gives no indication of the *rate* at which a system will move towards equilibrium.

12.3
HETEROGENEOUS EQUILIBRIA

Objectives

1 To extend the study of equilibria to those involving solids.

2 To introduce solubility product as a constant arising from a study of heterogeneous equilibria involving ions.

Timing

About two periods.

Suggested treatment

Most of the equilibria considered so far have been concerned with homogeneous systems in which only one phase is involved. In this section *heterogeneous* systems are dealt with, that is those in which the reactants or products occur in two or more phases. A common example of a heterogeneous equilibrium is a solid salt in contact with its ions in solution.

This type of system is used to introduce the idea of *solubility product*, which may be defined as the product of the ionic concentrations which exist at equilibrium in a saturated solution of a sparingly soluble ionic compound. The subject of solubility product is met again in Topic 15, which includes an experiment to find the solubility product of silver chloride (Experiment 15.3b).

Other heterogeneous equilibria

The rule that in heterogeneous equilibria the concentrations of all pure solid and pure liquid phases can be taken as constant can be illustrated by examples of other heterogeneous equilibria.

For example, for the system

$$CaCO_3(s) \rightleftharpoons CaO(s) + CO_2(g)$$

instead of writing

$$K_c = \frac{[CaO(s)]_{eqm}[CO_2(g)]_{eqm}}{[CaCO_3(s)]_{eqm}}$$

the expression $K_c = [CO_2(g)]_{eqm}$ is used. (The value of K_c in this expression is 3.4×10^{-5} mol dm^{-3} at 873 K.)

Similarly for this system,

$$K_p = p_{CO_2 eqm} \text{ and not } \frac{p_{CaO eqm} p_{CO_2 eqm}}{p_{CaCO_3 eqm}}$$

For this system $K_p = 2.4 \times 10^{-3}$ atm at 873 K, and 1.4 atm at 1173 K.

Other systems which could be considered briefly are:

$H_2O(l) \rightleftharpoons H_2O(g); K_p = p_{H_2O eqm}$

$NaNO_3(s) \rightleftharpoons NaNO_3(aq); K_c = [NaNO_3(aq)]_{eqm}$

$H_2O(g) + C(s) \rightleftharpoons H_2(g) + CO(g); K_p = \frac{p_{H_2 eqm} p_{CO eqm}}{p_{H_2O eqm}}$

$AgCl(s) \rightleftharpoons Ag^+(aq) + Cl^-(aq); K_c = [Ag^+(aq)]_{eqm}[Cl^-(aq)]_{eqm}$

$$Cu(s) + 2Ag^+(aq) \rightleftharpoons Cu^{2+}(aq) + 2Ag(s); \; K_c = \frac{[Cu^{2+}(aq)]_{eqm}}{[Ag^+(aq)]^2_{eqm}}$$

There is an opportunity here to point out that a similar convention is used for the concentration of solvents in dilute solutions, the most important example being the ionization of water.

$$H_2O(l) \rightleftharpoons H^+(aq) + OH^-(aq)$$

The change in the concentration of water as the degree of ionization varies with temperature is so small that it can be neglected. Hence [$H_2O(l)$] can be treated as constant and $[H^+(aq)]_{eqm}[OH^-(aq)]_{eqm} = K_c$. (This constant is usually given the special symbol K_w.)

The main point of introducing solubility product at this stage is to give an example of an equilibrium in which ions are involved. It is not intended that time should be devoted to a discussion of solubility products in general.

Suggestion for homework

Answering questions 12 and 13 at the end of this Topic in the *Students' book*.

Summary

At the end of this section students should:

    1 understand what is meant by a heterogeneous equilibrium;

    2 understand the meaning of solubility product, and realize that the term applies only to sparingly soluble electrolytes;

    3 know that in heterogeneous equilibria the concentration of all pure solid and pure liquid phases can be taken as constant.

12.4
ACID-BASE EQUILIBRIA

Objectives

1 To introduce the Lowry–Brønsted theory of acids and bases.
2 To introduce the ionization constant for water.
3 To consider the strengths of different acids and bases, and the effect of strength on pH of solution.
4 To study pH changes during acid-base reactions as shown by titration curves for strong and weak acids and bases.

Timing

About 6 or 7 periods.

Suggested treatment

For this treatment overhead projection transparency number 108 will be useful.

This part of the Topic is developed along fairly orthodox lines and has been dealt with in some detail in the *Students' book*. The text could be read by students before the subject matter is discussed in class.

The ground to be covered includes:

1 The ionization constant for water. Consideration of this leads naturally from the last part of the previous section.

2 The idea of an acid-base system as a competition for protons, acids being proton donors and bases proton acceptors.

3 The meaning of the word *strength* as applied to acids and bases.

4 The use of the pH meter. At this stage no attempt should be made to explain how the instrument works; this can be discussed if desired after consideration of redox potentials and the Nernst equation in Topic 15.

5 How to find the strengths of acids by measuring the pH of their solutions.

6 K_a, the dissociation constant of an acid.

7 Calculation of K_a from the pH of a solution of known concentration, and of pH of a solution of known concentration of an acid of known K_a.

The use of pH values in distinguishing between strong and weak acids can usefully be demonstrated; such a demonstration affords a good opportunity to describe how to use a pH meter before the students themselves use these instruments in Experiment 12.4.

TEACHER DEMONSTRATION
To distinguish strong and weak acids by pH measurements

The teacher will need:
pH meter and electrode
1.0M, 0.1M, 0.01M, and 0.001M solutions of ethanoic and hydrochloric acids
Further solutions of ethanoic acid of molarity 0.5M, 0.2M, 0.05M, and 0.02M, to enable a total of eight measurements to be made with this acid

Procedure

Measure the pH of the first set of solutions. Wash the electrodes between measurements and start from the most dilute solution to minimize the effect of contamination.

Discuss the values obtained in terms of dilution and possible increased ionization on dilution (see *Students' book*).

Now measure the pH of the second set of solutions and calculate values of K_a from this. For all but the highest and lowest concentrations, K_a should lie between 1.5 and 2.5×10^{-5}, and at the extremes of the concentration range studied it should not be far outside these limits. An example of this type of calculation for methanoic acid is given in the *Students' book*.

The reverse calculation, of pH from K_a, can conveniently be introduced at this stage. The *Students' book* contains such a problem. This problem can be worked out in a similar way to the calculation of the value of K_a for methanoic acid from the pH of a methanoic acid solution, but it involves a quadratic equation. It is worth while to point out to students that an approximation can simplify the equation greatly. This problem, and its solution, are as follows:

Calculate the pH of a 0.001M solution of glycine (aminoethanoic acid),

$NH_2CH_2CO_2H$ ($K_a = 1.7 \times 10^{-10}$ mol dm^{-3})

$$NH_2CH_2CO_2H(aq) \rightleftharpoons NH_2CH_2CO_2^-(aq) + H^+(aq)$$

$$K_a = \frac{[NH_2CH_2CO_2^-(aq)]_{eqm}[H^+(aq)]_{eqm}}{[NH_2CH_2CO_2H(aq)]_{eqm}}$$

Neglecting the hydrogen ions which arise from ionization of the water, since the concentration of these will be very small compared with the concentration of those from the acid, we can say that

$$[H^+(aq)]_{eqm} = [NH_2CH_2CO_2^-(aq)]_{eqm}$$

and $[NH_2CH_2CO_2H(aq)]_{eqm} = 0.001 - [H^+(aq)]_{eqm}$

$$\therefore \quad 1.7 \times 10^{-10} = \frac{[H^+(aq)]_{eqm}^2}{10^{-3} - [H^+(aq)]_{eqm}}$$

At this point, the quadratic can be worked out:

$$[H^+(aq)]_{eqm}^2 + (1.7 \times 10^{-10}[H^+(aq)]_{eqm}) - (1.7 \times 10^{-10} \times 10^{-3}) = 0$$

giving a value for the pH of 6.39.

Alternatively, an approximation can be made. Since $[H^+(aq)]_{eqm}$ will be very small for weak acids, we can omit the $1.7 \times 10^{-10}[H^+(aq)]_{eqm}$ term and simplify the calculation to

$[H^+(aq)]^2_{eqm} - 1.7 \times 10^{-10} + 10^{-3} = 0$

so that $[H^+(aq)]^2_{eqm} = 1.7 \times 10^{-13}$

and $[H^+(aq)]_{eqm} = \sqrt{(1.7 \times 10^{-13})} = 4.12 \times 10^{-7}$

from which pH $= -\lg 4.12 \times 10^{-7} = 6.39.$

For methanoic acid (a much stronger acid than glycine) in 0.01M solution, using $K_a = 1.82 \times 10^{-4}$ mol dm^{-3} (the value calculated on pages 29–30 of the *Students' book*) the simplified calculation gives

$$pH = -\lg\sqrt{(1.82 \times 10^{-4} \times 10^{-2})}$$
$$= 2.87$$

which agrees well with the value of 2.90 given there.

Although accurate results cannot be expected, the pH of the purest available water could be measured at some convenient stage. Discussion of the deviation from the expected value of 7 can be useful.

No treatment of strength of bases is recommended. Every weak base has a conjugate acid, and K_a values can cover all possibilities in this area.

Change of pH during acid-base titrations

This section can be completed by consideration of the changes in pH which take place during titration.

EXPERIMENT 12.4
An investigation of the change of pH during acid-base titration

Each student or pair of students will need:
Burette and burette stand
Pipette and pipette filler (or second burette and stand)
Beaker, 100 cm^3
pH meter and electrode
Magnetic stirrer, if possible
1.0M hydrochloric acid, 25 cm^3
1.0M ethanoic acid, 25 cm^3
1.0M sodium hydroxide solution, 40 cm^3
1.0M ammonia solution, 40 cm^3

Procedure

Full details of the experimental procedure are given in the *Students' book*. There are four different combinations of acid and base possible, namely

> strong acid and strong base
> strong acid and weak base
> weak acid and strong base
> weak acid and weak base

Time can be saved by allocating the various titrations to different students, so that each group does one titration. The results can be collected together and discussed at the end of the experiment.

Discussion

In the graphs of volume of alkali added against pH, it is the regions of rapid pH change that are of interest to us. It will be seen that in the strong acid/strong base titration there is a rapid change of pH from about 3 to about 10 due to a very small addition of base solution around the 'end-point'. This range is smaller in the strong acid/weak base titration and covers approximately pH 3–7. This is also the case in the weak acid/strong base titration but here the pH range of rapid change is approximately 7–11. For the weak acid/weak base system there is no marked horizontal portion in the curve and hence, no major region of rapid pH change.

This difference in the pH range at which equal volumes of solution react completely (25 cm^3 of each in this case) focuses attention on what exactly we mean by the end-point of a titration. In the examples under discussion, the solutions are all of the same molarity, and equal volumes should react together. The end-point therefore means the *equivalence-point* when the quantities of substance specified in the equation have reacted together. If this point can be made to correspond with a colour change in a suitable indicator we have a means of detecting the end-point easily and simply. Experience shows that choice of the correct indicator for use in acid/base titrations depends on the type of system used.

1 For strong acids with strong bases almost any indicator can be used.

2 For strong acids with weak bases the choice is restricted; methyl red or methyl orange are commonly used.

3 For weak acids with strong bases the choice is again restricted but indicators different from those in (2) must be used; phenolphthalein is a common choice.

4 For weak acids with weak bases it is very difficult to find a suitable indicator.

12.4 Acid-base equilibria

The suitability of indicators for different purposes was established by trial and error long before the idea of pH was introduced. From such information we might confidently guess that the colour change with methyl red or methyl orange occurs somewhere in the pH range 3–7, and for phenolphthalein somewhere in the pH range 7–11.

In cases where the use of indicators is unsuitable or impossible, a pH meter provides a simple method of following an acid–base titration.

From the titration curves it is obvious that the equivalence-point in a titration does not invariably correspond with the production of a neutral solution, which would be defined as one with pH 7. In the strong acid/strong base system a solution very near to this pH value is obtained. With a strong acid and a weak base the mixture is acid (low pH), and with a weak acid and a strong base it is alkaline (high pH). To understand the reasons for these differences we must look more closely at conditions when reacting quantities of acid and base are present in the three examples.

For the strong acid/strong base example, the reaction can be written

$$Na^+(aq) + OH^-(aq) + H^+(aq) + Cl^-(aq) \longrightarrow$$
$$H_2O(l) + Na^+(aq) + Cl^-(aq)$$

Here the $Na^+(aq)$ and $Cl^-(aq)$ ions have no very pronounced acidic or basic properties and the equilibrium established is

$$H^+(aq) + OH^-(aq) \rightleftharpoons H_2O(l)$$

for which, as we have seen, the pH is 7 when $[H^+(aq)]_{eqm} = [OH^-(aq)]_{eqm}$.

With the strong acid/weak base system we can represent the reaction as

$$NH_3(aq) + H^+(aq) + Cl^-(aq) \longrightarrow NH_4^+(aq) + Cl^-(aq)$$

Again the $Cl^-(aq)$ ion has no marked acidic or basic properties but the $NH_4^+(aq)$ ion is a not insignificant acid, which dissociates to $NH_3(aq)$ and $H^+(aq)$

$$NH_4^+(aq) \rightleftharpoons NH_3(aq) + H^+(aq)$$

For this equilibrium $K_a = 6 \times 10^{-10}$ mol dm^{-3} and from this we can calculate the pH of the solution obtained at equivalence-point in the titration. Here 25 cm^3 0.1M HCl(aq) has reacted with 25 cm^3 0.1M NH_3(aq). The result will be 50 cm^3 of 0.05M NH_4Cl(aq). Since K_a for the equilibrium shown above is small we shall not be seriously in error if we put $[NH_4^+(aq)]_{eqm} = 0.05$ mol dm^{-3}. Also, if we neglect the effect of the $OH^-(aq)$ ions from the ionization of the water (this will have a very small influence on the $H^+(aq)$ ion concentration), we can assume that

$$[NH_3(aq)]_{eqm} = [H^+(aq)]_{eqm}$$

Thus
$$K_a = \frac{[NH_3(aq)]_{eqm}[H^+(aq)]_{eqm}}{[NH_4^+(aq)]_{eqm}}$$

$$6 \times 10^{-10} = \frac{[H^+(aq)]^2_{eqm}}{0.05}$$

$\therefore$
$$[H^+(aq)]_{eqm} = \sqrt{(6 \times 10^{-10} \times 0.05)}$$
$$= 5.5 \times 10^{-6}$$

$$-\lg[H^+(aq)]_{eqm} = -(\overline{6}.74)$$
$$= -(-6 + 0.74)$$
$$= 5.26$$

$\therefore$ pH of solution at equivalence-point = 5.26

This pH should agree with the pH at the mid-point of the horizontal section of the titration curve.

For the weak acid/strong base titration the equation can be written

$$Na^+(aq) + OH^-(aq) + CH_3CO_2H(aq) \longrightarrow Na^+(aq) + CH_3CO_2^-(aq) + H_2O(l)$$

Neglecting $Na^+(aq)$ this reduces to

$$OH^-(aq) + CH_3CO_2H(aq) \rightleftharpoons CH_3CO_2^-(aq) + H_2O(l)$$

The ethanoate ion is a comparatively strong base and the equilibrium position is not entirely over to the right. A calculation similar to that given above shows that the pH of 0.05M sodium ethanoate solution is 8.75, which should agree with the mid-point of the horizontal portion of the graph obtained from the experiment.

For the ammonia/ethanoic acid titration the pH at equivalence-point will be the average of the two values given above, i.e. $\frac{5.26 + 8.75}{2} = 7.00$. The fact that this coincides with the neutral point is accidental. If methanoic acid is titrated with ammonia solution the pH at equivalence-point is 6.75. The value varies with the K_a values for the acid systems involved in the equilibria established at the equivalence-point.

It would be useful at this stage to add Full-range Indicator solution to solutions of the salts obtained in the acid–base reactions which have been studied ($NaCl$, NH_4Cl, CH_3CO_2Na, $CH_3CO_2NH_4$) and check that the above argument is reasonable.

Changes in pH value at the end-point of a titration

The origin of the rapid changes in pH at the end-point of the strong acid/strong base titration can be seen from a simple calculation of the pH value one drop (about $0.05\,cm^3$) before the equivalence-point, and one drop after the equivalence-point.

For one drop before the equivalence-point there will be $25.00\,cm^3$ 1.0M HCl(aq) present and $24.95\,cm^3$ 1.0M NaOH(aq), i.e. an excess of $0.05\,cm^3$ 1.0M HCl(aq) in $49.95\,cm^3$ (near enough to $50\,cm^3$) total solution

$$H^+(aq) \text{ ions per } 50\,cm^3 = \frac{0.05 \times 1.0}{1000}\,mol$$

$$\therefore [H^+(aq)] = \frac{0.05 \times 1.0 \times 1000}{1000 \times 50}\,mol\,dm^{-3}$$

$$= 10^{-3}\,mol\,dm^{-3}$$

$$\therefore \text{pH value} = 3$$

One drop after equivalence-point we have $0.05\,cm^3$ 1.0M NaOH(aq) in excess

$$OH^-(aq) \text{ ions per } 50\,cm^3 = \frac{0.05 \times 1.0}{1000}\,mol$$

$$\therefore [OH^-(aq)] = \frac{0.05 \times 1.0 \times 1000}{1000 \times 50} = 10^{-3}\,mol\,dm^{-3}$$

but $\quad [H^+(aq)][OH^-(aq)] = K_w = 10^{-14}$

$$\therefore [H^+(aq)] = \frac{10^{-14}}{10^{-3}} = 10^{-11}$$

$$\therefore \text{pH value} = 11$$

Thus there is a change in pH value of 8 units during the addition of two drops of 1.0M NaOH(aq) at the end-point of this titration.

For the titration of a strong acid with a weak base the corresponding rapid change in pH extends over the range 3–7 only, and for a weak acid and a strong base over the range 7–11. For a weak acid/weak base system there is no region where the pH change is of this magnitude.

Suggestions for homework

Calculations on pH values of weak acid solutions of different molarities.

Summary

At the end of this section students should:
1 be able to discuss acid–base theory on a qualitative basis;
2 be able to identify species which are acting as acids and bases in a given system;
3 know what is meant by pH and K_a;
4 be able to discuss the pH changes that take place during acid–base reactions, and relate these to the K_a values of the acids concerned.

12.5
BUFFER SOLUTIONS AND INDICATORS
Objectives

1 To introduce the idea of a buffer solution.
2 To develop an equation to use for calculations involving pH of buffer solutions.
3 To discuss the simple theory of indicators and to measure K_a for an indicator and for a weak acid.

Timing

About four periods.

Suggested treatment

For this treatment overhead projection transparency number 109 will be useful.

The subjects of buffer solutions and indicators are dealt with in some detail in the *Students' book* and this might be read by students beforehand. Little more needs to be added here. If a pH meter is available the effect of addition of acid and alkali to a buffer solution and to a non-buffer solution of approximately the same pH could be shown. Also a comparison of calculated and measured pH values for a given buffer solution would be of interest. In Experiment 12.5 students determine K_a for an indicator and for a weak acid.

EXPERIMENT 12.5
To measure K_a for (a) an indicator and (b) a weak acid

Each student or pair of students will need:
18 test-tubes, 125 × 16 mm (or equivalent size specimen tubes)
1 rack to hold the test-tubes in pairs one behind the other so that the colour can be seen through each pair of tubes
1 teat pipette to deliver approximately 0.5 cm^3
1 measuring cylinder (25 cm^3)

Access to:
Bromophenol blue solution (0.1 g dissolved in 20 cm^3 ethanol and then made up to 100 cm^3 with water)
Concentrated hydrochloric acid
Approximately 4M sodium hydroxide solution

and communal burettes containing:
0.02M methanoic acid (made up approximately from solution from suppliers and then standardized)
0.02M sodium methanoate (1.36 g HCO$_2$Na per dm^3)
0.02M benzoic acid (2.44 g C$_6$H$_5$CO$_2$H per dm^3. Make up in warm water and allow to cool)
0.02M sodium benzoate (2.88 g C$_6$H$_5$CO$_2$Na per dm^3)

Procedure

Students make up solution X which contains bromophenol blue indicator in the HIn form by adding one drop of concentrated hydrochloric acid to 5 cm^3 of indicator solution. They also make up solution Y in which the indicator is in the In$^-$ form by adding one drop of 4M sodium hydroxide solution to 5 cm^3 of indicator solution.

They then set up nine pairs of tubes so that the colour *when looking through both tubes* is that due to the indicator with the ratios of the [HIn]:[In$^-$] form as follows: 1:9, 2:8, 3:7, and so on.

Students then find which pair of tubes matches most closely the indicator colour produced by mixing 5 cm^3 of 0.02M sodium methanoate, 5 cm^3 of 0.02M methanoic acid, and 10 drops of bromophenol blue solution. The answers to the questions at the end of the instructions in the *Students' book* should be as follows.

1 The pair of tubes with the [HIn]:[In$^-$] ratio = 6:4 will probably match the colour of the solution best. This is assumed in the following answers.

2 The pH of the solution is given by

$$pH = -\lg K_a - \lg \frac{[\text{acid}]_{\text{eqm}}}{[\text{base}]_{\text{eqm}}}$$

Assuming all the methanoate ions come from the sodium methanoate,

$$pH = -\lg(2 \times 10^{-4}) - \lg \frac{0.01}{0.01}$$

$$= 3.7$$

3 The ratio of $[\text{HIn}]_{\text{eqm}} : [\text{In}^-]_{\text{eqm}}$ at this pH is $6:4$.

4 K_a for bromophenol blue is given by

$$pH = -\lg K_a - \lg \frac{[\text{HIn(aq)}]_{\text{eqm}}}{[\text{In}^-(\text{aq})]_{\text{eqm}}}$$

$$3.7 = -\lg K_a - \lg \frac{6}{4}$$

$$\lg K_a = -3.88$$

$$K_a = 1.3 \times 10^{-4} \text{ mol dm}^{-3}$$

Students then measure K_a for benzoic acid by finding which pair of tubes matches most closely the indicator colour produced by mixing 5 cm^3 of 0.02M sodium benzoate, 5 cm^3 of 0.02M benzoic acid, and 10 drops of bromophenol blue. Answers to the questions given after this experiment should be as follows.

1 The pair of tubes with the $[\text{HIn}]_{\text{eqm}} : [\text{In}^-]_{\text{eqm}}$ ratio $= 3:7$ will probably match the colour of this solution best. This is assumed in the following answers.

2 The $[\text{HIn}]_{\text{eqm}} : [\text{In}^-]_{\text{eqm}}$ ratio is 0.429.

3 The pH of the mixture is given by:

$$pH = -\lg K_a - \lg \frac{[\text{HIn(aq)}]_{\text{eqm}}}{[\text{In}^-(\text{aq})]_{\text{eqm}}}$$

$$= -\lg(1.3 \times 10^{-4}) - \lg 0.429$$

$$= 4.26$$

4 K_a for benzoic acid is given by

$$pH = -\lg K_a - \lg \frac{[\text{acid}]_{\text{eqm}}}{[\text{base}]_{\text{eqm}}}$$

$$4.26 = -\lg K_a - \lg \frac{0.01}{0.01}$$

$$\lg K_a = -4.26$$

$$K_a = 5.5 \times 10^{-5} \text{ mol dm}^{-3}$$

The accuracy of these results depends on the accuracy with which the solutions have been made up, and also on the size of the gap between the sets of tubes. It would be instructive for students to calculate the value of K_a in each case using the $[\text{HIn}]_{\text{eqm}} : [\text{In}^-]_{\text{eqm}}$ ratio for the pair of tubes *next* to the pair that they used.

Formulae for indicators

Students will probably be curious about these. A few examples are given below.

Methyl orange

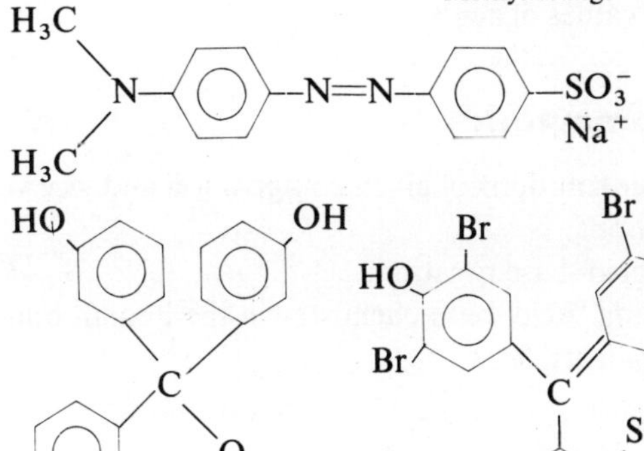

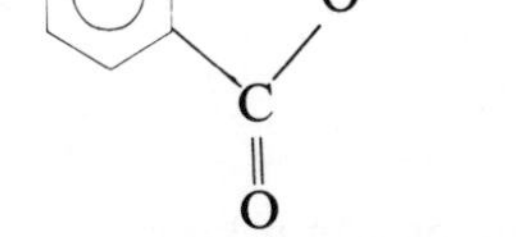

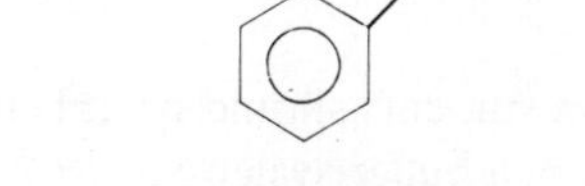

Phenolphthalein

Bromophenol blue

Enquiries about litmus should be answered by referring to an article by H. G. Andrew on the subject in *School Science Review*, 153, *44*, p. 338 (1963).

pK_a values

The *Students' book* contains a brief explanation of the use of this terminology.

Background reading

Finally in this section of the *Students' book* there is a passage of Background reading entitled 'Acid–base chemistry in the human body'.

At the end of this piece of Background reading the student is set a question. It can be solved in the following way.

The original blood has pH = 7.25 and p_{CO_2} = 42 mmHg. For these values the nomogram shows that $[HCO_3^-(aq)]_{eqm}$ = 17 mmol dm^{-3}. If the blood is to be stabilized at pH = 7.45 and p_{CO_2} = 30 mmHg, the nomogram shows that $[HCO_3^-(aq)]_{eqm}$ must be 19 mmol dm^{-3}. An extra 2 mmol dm^{-3} must therefore be added to the blood; 10 dm^3 will require 20 mmol to be added.

The problem can also be solved using the acid–base equations.

Supporting material

The *Book of data* contains tables of solubility products, the formulation of buffer solutions, and the K_a and pK_a values of acids.

Suggestions for homework

Calculating pH values for buffer solutions of given composition and vice versa.
Looking up uses of buffer solutions.
Choosing indicators for given acid–base titrations.
Reading the Background reading 'Acid–base chemistry in the human body', at the end of this section in the *Students' book*.

Summary

At the end of this section students should be able to:

1 explain the action of a buffer system;

2 calculate the pH of buffer solutions when composition is known and vice versa;

3 understand the elementary theory of indicators;

4 understand what is meant by the pK_a value of an acid;

5 appreciate some of the aspects of acid–base chemistry that concern the human body.

12.6
EQUILIBRIUM AND ZERO TOTAL ENTROPY CHANGE
Objectives

1 To extend the relationship

$$\Delta S_{\text{total}} = 0$$

for a system in equilibrium, developed earlier (Topic 10.1) as a means of discovering whether or not a given change can proceed, using as examples the freezing of water and the reaction

$$N_2(g) + 3H_2(g) \longrightarrow 2NH_3(g)$$

2 To calculate the value of K_p for the equilibrium

$$N_2(g) + 3H_2(g) \rightleftharpoons 2NH_3(g)$$

3 To introduce the concept of standard free energy change ($\Delta G^\ominus$) of a reaction as arising from the relationship

$$\Delta G^\ominus = -T\Delta S_{\text{total}}$$

4 To develop the relationship

$$\Delta G^\ominus = -LkT \ln K_p$$

Timing

Four periods should be ample. This section needs to be taken fairly slowly to give opportunities for questions and discussion. It will be helpful if students read the *Students' book* section both before and after it is dealt with in class.

Suggested treatment

As this section is concerned solely with imparting information, and does not include experimental work, teachers will probably find it most satisfactory to follow the given sequence.

Students will probably want to know what is meant by 'kinetic reasons' at the beginning of the discussion of the 'entropy balance sheet' for the ammonia equilibrium (page 52). This is dealt with later (Topic 14). Here it will be sufficient for the teacher to point out that the fact that the value of ΔS_{total} is positive for a suggested change is not a guarantee that the change will take place under the conditions assumed (here 298 K and 760 mmHg pressure). Almost all reactions proceed in successive stages, some of which need very considerable amounts of energy to be supplied, while others are accompanied by the evolution of energy. A stage which is markedly endothermic constitutes a barrier to the overall change taking place and under standard conditions sufficient energy may not be available for this stage to proceed.

The high calculated value for K_p (page 53) in the ammonia equilibrium under standard conditions (about 10^5 atm^{-2}) is thus not an indication that the synthesis of ammonia can be achieved successfully under these conditions. If it was so then the release of hydrogen into a laboratory should produce a reaction and the smell of ammonia! The Haber process for the manufacture of ammonia is carried out at high temperatures and pressures. As was shown earlier in this Topic (table 12.4) the value of K_p decreases rapidly as the temperature is raised. Hence there is a need to conduct the process at as high a pressure as possible and at a temperature as low as possible, consistent with the reaction rate being sufficiently high to enable reasonable yields of ammonia to be achieved. In practice this entails temperatures in the region of 300–400 °C.

After the ammonia equilibrium has been discussed in class students could be asked to carry out a similar exercise, under supervision, this time using the balance sheet to obtain a value of K_p for the equilibrium

$$C(\text{graphite}) + 2H_2(g) \rightleftharpoons CH_4(g)$$

The results are as follows.

Event	Entropy change/J K^{-1}
lose x mole C	$-5.7x$
lose $2x$ mole H$_2$	$-2(130.6)x$
gain x mole CH$_4$	$+186.2x$
total so far	$-80.7x$
give 74.8x kJ to surroundings at 298 K (74 800x/298)	$+251.0x$
reaction with reactants at standard state	$+170.3x$

$$\ln K_p = \ln\left[\frac{p_{CH_4\text{eqm}}}{p^2_{H_2\text{eqm}}}\right] = \frac{170.3}{Lk} = \frac{170.3}{6 \times 10^{23} \times 1.4 \times 10^{-23}}$$
$$= 20.3$$

Therefore $K_p = \exp(20.3) \approx 10^9$ atm^{-1}

(The partial pressure of C(graphite) is constant, and is therefore incorporated into the equilibrium constant, K_p.)

This result shows that, on purely energetic considerations, carbon and hydrogen should be almost completely converted to methane at room temperature and pressure. In fact the conversion has not been achieved and no catalyst has yet been found that can enable it to be done. Although the reaction is feasible it does not take place. Hence there must be other factors which determine whether or not a reaction will proceed, besides those concerned solely with overall energy transfers.

Supporting homework

Answering question 21 at the end of the Topic in the *Students' book*.

Summary

At the end of this section students should:

1 have a better understanding of the role that entropy changes play in determining whether or not a change can proceed (that is, is feasible);

2 be able to draw up an entropy 'balance sheet' and use it to calculate K_p for a reaction;

3 have been introduced to the concept of standard free energy change. The use of this will be developed further in Topic 15.

ANSWERS TO PROBLEMS IN THE *STUDENTS' BOOK*

(A suggested mark allocation is given in brackets after each answer.)

1 11 g of ethyl ethanoate $= \frac{1}{8}$ mole (2)

18 g of water	$= 1$ mole	(2)
Ethanoic acid formed	$\equiv (106 - 18) \text{ cm}^3 \text{M NaOH}$	(2)
	$= 0.088$ mole	(2)
Ethanol formed	$= 0.088$ mole	(2)
$[CH_3CO_2C_2H_5(l)]_{eqm}$	$= 0.125 - 0.088$	(2)
$[H_2O(l)]_{eqm}$	$= 1 - 0.088$	(2)

$$K_c = \frac{[CH_3CO_2H(l)]_{eqm}[C_2H_5OH(l)]_{eqm}}{[CH_3CO_2C_2H_5(l)]_{eqm}[H_2O(l)]_{eqm}} \quad (2)$$

$$= \frac{(0.088)^2}{0.037 \times 0.912} \quad (2)$$

$$= 0.23 \quad (2)$$

Total 20 marks

Topic 12 Equilibria: gaseous and ionic

2a i 0.011 mole pentene (1)
ii 0.001 mole ethanoic acid (1)

b $K_c = \dfrac{[CH_3CO_2C_5H_{11}]_{eqm}}{[C_5H_{10}]_{eqm}[CH_3CO_2H]_{eqm}}$ (1)

c $[C_5H_{10}]_{eqm} = 0.011 \times \dfrac{10}{6}\,\text{mol dm}^{-3}$ (1)

$[CH_3CO_2H]_{eqm} = 0.001 \times \dfrac{10}{6}\,\text{mol dm}^{-3}$ (1)

$[CH_3CO_2C_5H_{11}]_{eqm} = 0.009 \times \dfrac{10}{6}\,\text{mol dm}^{-3}$ (1)

d $K_c = \dfrac{0.009 \times \dfrac{10}{6}}{0.011 \times \dfrac{10}{6} \times 0.001 \times \dfrac{10}{6}}\,\text{dm}^3\,\text{mol}^{-1}$ (2)

$= 491\,\text{dm}^3\,\text{mol}^{-1}$ (2)

Total 10 marks

3 $K_c = \dfrac{[NO_2]^2}{[N_2O_4]}$ (1)

$= \dfrac{0.0014^2}{0.13}\,\text{mol dm}^{-3}$ (2)

$= 1.5 \times 10^{-5}\,\text{mol dm}^{-3}$ (2)

Total 5 marks

4a $\dfrac{2.085 \times 2}{208.5} = 0.02\,\text{mol dm}^{-3}$ (2)

b i $[PCl_5(g)]_{eqm} = 0.02 - x\,\text{mol dm}^{-3}$ (1)
ii $[PCl_3(g)]_{eqm} = x\,\text{mol dm}^{-3}$ (1)

c $K_c = \dfrac{[PCl_3(g)]_{eqm}[Cl_2(g)]_{eqm}}{[PCl_5(g)]_{eqm}}$ (2)

d $0.19 = \dfrac{x^2}{(0.02 - x)}$ (2)

$x = 0.018\,\text{mol dm}^{-3}$ (absurd root is -0.208) (5)

e $[PCl_5]_{eqm} = 0.002 \text{ mol dm}^{-3}$ (1)
$[PCl_3]_{eqm} = [Cl_2]_{eqm} = 0.018 \text{ mol dm}^{-3}$ (1)
Total 15 marks

5 Initial concentration of propanone = 0.05 mol dm^{-3} (1)
Initial concentration of HCN = 0.1 mol dm^{-3} (1)
Let $[product]_{eqm} = x \text{ mol dm}^{-3}$

$$\frac{x}{(0.05-x)(0.1-x)} = 32.8$$ (2)

$x = 0.034 \text{ mol dm}^{-3}$ (5)
Final volume of solution = 200 cm^3 (1)

containing $\frac{200 \times 0.034}{1000}$ mole product (2)

The relative molecular mass of the product = 85 (1)

∴ Mass of product = $\frac{200 \times 0.034 \times 85}{1000}$ g (1)

= 0.58 g (1)
Total 15 marks

6a 2 moles of $S_2O_3^{2-}$(aq) react with 1 mole of I_2(aq) (1)

6.4 cm^3 of 0.1M thiosulphate solution contain $\frac{6.4 \times 0.1}{1000}$ moles of $S_2O_3^{2-}$(aq) (1)

So the amount of I_2(g) present at equilibrium, in moles, was
$\frac{6.4 \times 0.1}{1000 \times 2} = 0.00032 \text{ mol } I_2$(g) (1)

b 0.00032 mol H_2(g) present at equilibrium (1)
c 2 moles of HI give 1 mole of I_2 (1)
∴ $2 \times 0.00032 = 0.00064$ mole of HI was converted to I_2 (1)
∴ $0.0023 - 0.00064 = 0.00166$ mole of HI was present at equilibrium (1)

d $K_c = \frac{0.00032^2}{0.00166^2}$ (2)

= 0.037 (1)
Total 10 marks

7 The relative molecular mass of $C_2H_5CO_2C_2H_5 = 102$ (1)

$\therefore \dfrac{80}{102} = 0.78$ mole $C_2H_5CO_2C_2H_5(l)$ is present at equilibrium (1)

The relative molecular mass of $C_2H_5CO_2H = 74$ (1)

$\dfrac{60}{74} = 0.81$ mole

$\therefore 0.81 - 0.78 = 0.03$ mole of $C_2H_5CO_2H(l)$ is present at equilibrium (2)

Let x g be the mass of $C_2H_5OH(l)$ needed
The relative molecular mass of $C_2H_5OH = 46$ (1)

$\therefore \dfrac{x}{46} - 0.78$ mole $C_2H_5OH(l)$ is present at equilibrium (2)

$K_c = \dfrac{0.78^2}{0.03\left(\dfrac{x}{46} - 0.78\right)} = 7.5$ (2)

$x = 160$ g (5)

Total 15 marks

8 $K_p = \dfrac{p_{SO_3 eqm}^2}{p_{SO_2 eqm}^2 \times p_{O_2 eqm}}$ (2)

At equilibrium $1\tfrac{1}{3}$ mole $SO_2 + \tfrac{2}{3}$ mole $O_2 \rightleftharpoons \tfrac{2}{3}$ mole SO_3 (2)

$p_{SO_3 eqm} = \dfrac{\tfrac{2}{3}}{2\tfrac{2}{3}} \times 9$ atm (1)

$p_{SO_2 eqm} = \dfrac{1\tfrac{1}{3}}{2\tfrac{2}{3}} \times 9$ atm (1)

$p_{O_2 eqm} = \dfrac{\tfrac{2}{3}}{2\tfrac{2}{3}} \times 9$ atm (1)

Hence $K_p = \dfrac{1}{9}$ atm^{-1} (3)

(Allow 1 out of 3 for correct answer; 2 out of 3 for units, atm^{-1}.)

Total 10 marks

9a $K_p = \dfrac{p_{NOeqm}^2 \times p_{O_2eqm}}{p_{NO_2eqm}^2}$ (2)

$p_{NOeqm} = \dfrac{0.04}{1.02} \times 0.2 \text{ atm}$ (2)

$p_{O_2eqm} = \dfrac{0.02}{1.02} \times 0.2 \text{ atm}$ (2)

$p_{NO_2eqm} = \dfrac{0.96}{1.02} \times 0.2 \text{ atm}$ (2)

$K_p = \dfrac{\left(\dfrac{0.04 \times 0.2}{1.02}\right)^2 \times \left(\dfrac{0.02 \times 0.2}{1.02}\right)}{\left(\dfrac{0.96 \times 0.2}{1.02}\right)^2}$ (1)

$= \dfrac{0.04^2 \times 0.02 \times 0.2}{1.02 \times 0.96^2}$

$= 6.8 \times 10^{-6} \text{ atm}$ (1)

b $M_{av} = 46 \times \dfrac{0.96}{1.02} + 30 \times \dfrac{0.04}{1.02} + 32 \times \dfrac{0.02}{1.02}$ (3)

$= 45.1$ (2)

Total 15 marks

10a $K_p = \dfrac{p_{COeqm} \times p_{H_2Oeqm}}{p_{CO_2eqm} \times p_{H_2eqm}}$ (2)

$3.8 \times 10^{-5} = \dfrac{x^2}{(1-x)^2}$ where x = moles of CO (3)

x being small, $x^2 = 3.8 \times 10^{-5}$
$x = 6.2 \times 10^{-3}$ mole (2)

b 6.2×10^{-3} (2)
It is suggested that no marks be given if this answer is reached by a complete re-working of the problem. An application of Le Châtelier's principle shows that neither the forward nor the reverse reaction results in a decrease in volume, so a change in overall pressure does not alter the position of equilibrium. (3)

c Less. (1)
A smaller equilibrium constant indicates that the concentration of the products of the forward reaction (righthand side of the equation) will be less. (3)

d Endothermic. (1)
Application of Le Châtelier's principle indicates that the higher temperature favours the endothermic reaction. Since K_p is higher at 373 K than it is at 298 K, the forward reaction is favoured by the higher temperature. Therefore the forward reaction is endothermic. (3)

Total 20 marks

11a If the mole fraction of CO is x,
$28x + 44(1 - x) = 36$, (2)
$x = \frac{1}{2}$ (1)

b $K_p = \dfrac{p^2_{\text{COeqm}}}{p_{\text{CO}_2\text{eqm}}}$ (3)

$= \dfrac{(\frac{1}{2} \times 12)^2}{\frac{1}{2} \times 12}$ (2)

$= 6 \text{ atm}$ (1)

c If the mole fraction of CO is y

$K_p = 6 \text{ atm} = \dfrac{(y \times 2)^2}{2(1 - y)}$ (3)

$\therefore y^2 + 3y - 3 = 0$ (1)
$y = 0.8$ (2)

Total 15 marks

12 From the *Book of data*, the solubility of strontium carbonate is 7.38×10^{-6} moles per 100 g water at 25 °C, that is, 7.38×10^{-5} mol dm^{-3}. (2)
K_{sp} for $SrCO_3 = (7.38 \times 10^{-5})^2$ at 25 °C (2)
In the final solution $[CO_3^{2-}]_{\text{eqm}} = 0.5$ mol dm^{-3} (2)

$[Sr^{2+}]_{\text{eqm}} = \dfrac{(7.38 \times 10^{-5})^2}{0.5}$ (2)

$= 1.09 \times 10^{-8}$ mol dm^{-3} (2)

Total 10 marks

13 $[Ag^+(aq)][Cl^-(aq)] = 0.0005 \times 0.0005$
$= 2.5 \times 10^{-7}$ mol^{-2} dm^{-6} (1)
K_{sp}AgCl at 25 °C $= 2 \times 10^{-10}$ mol^2 dm^{-6} (1)
This is less than 2.5×10^{-7} (1)
$\therefore$ silver chloride is precipitated. (1)

b $[Ca^{2+}(aq)][CO_3^{2-}(aq)] = 2.5 \times 10^{-7} \, mol^2 \, dm^{-6}$ (1)
$K_{sp}CaCO_3 = 5 \times 10^{-9} \, mol^2 \, dm^{-6}$ (1)
This is less than 2.5×10^{-7} (1)
∴ calcium carbonate is precipitated. (1)

c $[Ag^+(aq)][BrO_3^-(aq)] = 2.5 \times 10^{-7} \, mol^2 \, dm^{-6}$ (1)
$K_{sp}AgBrO_3 = 6 \times 10^{-5} \, mol^2 \, dm^{-6}$ (1)
This is greater than 2.5×10^{-7} (1)
∴ no precipitation takes place. (1)

d $[Mg^{2+}(aq)][OH^-(aq)]^2 = 1.25 \times 10^{-10} \, mol^3 \, dm^{-9}$ (1)
$K_{sp}Mg(OH)_2 = 2 \times 10^{-11} \, mol^3 \, dm^{-9}$ (1)
This is less than 1.25×10^{-10} (1)
∴ magnesium hydroxide is precipitated. (1)

Total 16 marks

14a $[H^+(aq)]_{eqm} = 2 \times 10^{-1} \, mol \, dm^{-3}$ (1)
$pH = -\lg 2 \times 10^{-1}$ (1)
$pH = 0.7$ (1)

Total 3 marks

(In answers to this, and other parts of question 14, it is likely that many students will not show the steps by which they reached the pH value. Full marks may be given for the correct answer alone, on the understanding that no marks can be given for an incorrect answer alone. The marking scheme is structured so that credit can be given for partially correct answers in which the working *is* shown.)

b $[OH^-(aq)]_{eqm} = 2 \times 10^{-1} \, mol \, dm^{-3}$ (1)
$[H^+(aq)]_{eqm} = 10^{-14} \div (2 \times 10^{-1})$ (1)
$pH = -\lg 5 \times 10^{-14}$ (1)
$pH = 13.30$ (1)

Total 4 marks

c $[H^+(aq)]_{eqm} = 1.25 \times 10^{-1} \, mol \, dm^{-3}$ (1)
$pH = -\lg 1.25 \times 10^{-1}$ (1)
$pH = 0.90$ (1)

Total 3 marks

d Excess of acid ≡ $50 \, cm^3$ 0.1M in a total volume of $100 \, cm^3$ (1)
Concentration of acid = 0.05M (1)
$[H^+(aq)]_{eqm} = 5 \times 10^{-2} \, mol \, dm^{-3}$ (1)
$pH = -\lg 5 \times 10^{-2}$ (1)
$pH = 1.30$ (1)

Total 5 marks

e $CH_2BrCO_2H(aq) \rightleftharpoons CH_2BrCO_2^-(aq) + H^+(aq)$ (1)

$$K_a = \frac{[CH_2BrCO_2^-(aq)]_{eqm}[H^+(aq)]_{eqm}}{[CH_2BrCO_2H(aq)]_{eqm}}$$ (1)

$= 1.35 \times 10^{-3}$

Let $[H^+(aq)]_{eqm} = x$

Then $\frac{x^2}{0.1-x} = 1.35 \times 10^{-3}$ (3)

If x is small $0.1 - x \approx 0.1$
$\therefore x^2 = 1.35 \times 10^{-3} \times 10^{-1}$
$\quad = 1.35 \times 10^{-4}$ (1)
$\therefore x = 1.16 \times 10^{-2} \, mol\, dm^{-3}$ (1)
$\therefore pH = -\lg 1.16 \times 10^{-2} = 1.94$ (1)

Total 8 marks

15 For $HCO_2H(aq) \rightleftharpoons HCO_2^-(aq) + H^+(aq)$
$K_a = 1.6 \times 10^{-4}$
Neglecting the hydrogen ions which arise from ionization of the water,
$\quad [HCO_2^-(aq)]_{eqm} = [H^+(aq)]_{eqm}$
$\quad [HCO_2H(aq)]_{eqm} = 0.01 - [HCO_2^-(aq)]_{eqm}$

$\therefore \frac{[HCO_2^-(aq)]_{eqm}^2}{0.01 - [HCO_2^-(aq)]_{eqm}} = 2 \times 10^{-4}$ (1)

Ignoring $[HCO_2^-(aq)]_{eqm}$ with respect to 0.01:

$\frac{[HCO_2^-(aq)]_{eqm}^2}{0.01} = 1.6 \times 10^{-4}$ (1)

$[HCO_2^-(aq)]_{eqm} = 1.3 \times 10^{-3} \, mol\, dm^{-3}$ (1)

Total 3 marks

16 For $HA(aq) \rightleftharpoons H^+(aq) + A^-(aq)$

$$K_a = \frac{[H^+(aq)]_{eqm}[A^-(aq)]_{eqm}}{[HA(aq)]_{eqm}}$$

and $[A^-(aq)]_{eqm} = [H^+(aq)]_{eqm} = 1.3 \times 10^{-3} \, mol\, dm^{-3}$
$\therefore [HA(aq)]_{eqm} = [0.1 - (1.3 \times 10^{-3})] \, mol\, dm^{-3}$

$\therefore K_a = \frac{(1.3 \times 10^{-3})^2}{0.1 - (1.3 \times 10^{-3})}$ (1)

Ignoring 1.3×10^{-3} with respect to 0.1,

$$K_a = \frac{(1.3 \times 10^{-3})^2}{0.1}$$ (1)

$$= 1.7 \times 10^{-5}$$ (1)

Total 3 marks

17 $pH = -\lg [H^+(aq)]_{eqm} = 5.1$
∴ $[H^+(aq)]_{eqm} = 7.94 \times 10^{-6}$ (1)
For $HA(aq) \rightleftharpoons H^+(aq) + A^-(aq)$
$[H^+(aq)]_{eqm} = [A^-(aq)]_{eqm} = 7.94 \times 10^{-6} \, mol \, dm^{-3}$
$[HA(aq)]_{eqm} = [0.1 - (7.94 \times 10^{-6})] \, mol \, dm^{-3}$

∴ $K_a = \dfrac{[H^+(aq)]_{eqm}[A^-(aq)]_{eqm}}{[HA(aq)]_{eqm}}$

$ = \dfrac{(7.94 \times 10^{-6})^2}{0.1 - (7.94 \times 10^{-6})}$ (1)

Ignoring 7.94×10^{-6} with respect to 0.1

$K_a = \dfrac{(7.94 \times 10^{-6})^2}{0.1}$ (1)

$= 6.3 \times 10^{-10}$ (1)

Total 4 marks

18 For the equilibrium
$C_6H_5NH_3^+(aq) \rightleftharpoons H^+(aq) + C_6H_5NH_2(aq)$
$[C_6H_5NH_2(aq)]_{eqm} = [H^+(aq)]_{eqm}$
$[C_6H_5NH_3^+(aq)]_{eqm} = 0.001 - [H^+(aq)]_{eqm}$

∴ $\dfrac{[H^+(aq)]_{eqm}^2}{0.001 - [H^+(aq)]_{eqm}} = 2 \times 10^{-5}$ (1)

Ignoring $[H^+(aq)]_{eqm}$ with respect to 0.001
$[H^+(aq)]_{eqm} = 1.41 \times 10^{-4} \, mol \, dm^{-3}$ (1)
$pH = -\lg [H^+(aq)]_{eqm} = -(-3.9)$
$= 3.9$ (1)

Total 3 marks

19 K_a for propanoic acid $= 1.3 \times 10^{-5}$
$[acid]_{eqm} = 0.1 \text{ mol dm}^{-3}$
$[base]_{eqm} = 0.05 \text{ mol dm}^{-3}$

$$pH = -\lg(1.3 \times 10^{-5}) - \lg\frac{0.1}{0.05} \quad (1)$$

$$= 4.89 - 0.301$$
$$= 4.58 \quad (2)$$

Total 3 marks

20 K_a for ethanoic acid $= 1.7 \times 10^{-5}$

a $4.7 = -\lg(1.7 \times 10^{-5}) - \lg\left(\dfrac{[CH_3CO_2H]_{eqm}}{[CH_3CO_2^-]_{eqm}}\right) \quad (1)$

$4.7 = \quad 4.77 \quad -\lg\left(\dfrac{[CH_3CO_2H]_{eqm}}{[CH_3CO_2^-]_{eqm}}\right)$

$\lg\left(\dfrac{[CH_3CO_2H]_{eqm}}{[CH_3CO_2^-]_{eqm}}\right) = 0.07$

$\dfrac{[CH_3CO_2H]_{eqm}}{[CH_3CO_2^-]_{eqm}} = 1.17 \quad (2)$

The solutions must therefore be mixed in the proportions, 1.17 volumes of 0.1M ethanoic acid to 1 volume of 0.1M sodium ethanoate. (1)

b $4.4 = -\lg(1.7 \times 10^{-5}) - \lg\left(\dfrac{[CH_3CO_2H]_{eqm}}{[CH_3CO_2^-]_{eqm}}\right) \quad (1)$

$4.4 = 4.77 - \lg\left(\dfrac{[CH_3CO_2H]_{eqm}}{[CH_3CO_2^-]_{eqm}}\right)$

$\lg\left(\dfrac{[CH_3CO_2H]_{eqm}}{[CH_3CO_2^-]_{eqm}}\right) = 0.37$

$\dfrac{[CH_3CO_2H]_{eqm}}{[CH_3CO_2^-]_{eqm}} = 2.34 \quad (2)$

The solutions must therefore be mixed in the proportions, 2.34 volumes of 0.1M ethanoic acid to 1 volume of 0.1M sodium ethanoate. (1)

Total 8 marks

21a

Event	Entropy change/J K^{-1}	
lose $2x$ mole SO_2	$-2(248.1)x$	
lose x mole O_2	$-205.0x$	
gain $2x$ mole SO_3	$+2(256.1)x$	
total so far	$-189.0x$	(4)
give $197.0x$ kJ to surroundings at 298 K ($197\,000x/298$)	$+661.1x$	(1)
reaction with reactants at standard rate	$+471.2x$	(1)

b $\ln K_p = \dfrac{p^2_{SO_2\text{eqm}}}{p^2_{SO_2\text{eqm}}} = 56.2$

$K_p = \exp(56.2) \approx 10^{24}$ (4)

Total 10 marks

TOPIC 13
Carbon compounds with acidic and basic properties

OBJECTIVES

1 To apply the principles learnt in Topics 9 and 11 to the reactions of carboxylic acids, their derivatives, and amines and amino acids.
2 To learn about the reactions of carboxylic acids, their derivatives, and amines and amino acids.
3 To apply the principles learnt about hydrogen bonding and acid-base properties to the study of organic compounds.
4 To develop skill in the techniques of organic chemistry such as chromatography.
5 To learn about the catalytic properties of enzymes and become aware of their importance in biological processes.
6 To provide information about the industrial and social importance of selected compounds.

CONTENT

13.1 Carboxylic acids. Experiments with ethanoic acid; laboratory preparation of methyl benzoate; reactions of carboxylic acids. Background reading:1 'Naturally occurring carboxylic acids'.
13.2 Carboxylic acid derivatives. Experiments with carboxylic acid derivatives; laboratory preparation of cholesteryl benzoate 'liquid crystals'; reactions of carboxylic acid derivatives. Background reading:2 'Liquid crystal displays'.
13.3 Amines. Experiments with amines; laboratory preparations of 2-ethanoylaminobenzoic acid and of nylon; reactions of amines.
13.4 Amino acids and proteins. The twenty most important amino acids; experimental investigation of protein materials; chromatographic separation of amino acids. Background reading:3 'The chemical and structural investigation of proteins'.
13.5 Enzymes. Investigation of the enzyme-catalysed hydrolysis of urea; properties of enzymes. Background reading:4 'The industrial uses of enzymes', :5 'Enzymes in medicine'.
13.6 Survey of reactions in Topic 13.

TIMING

This Topic should take three to four weeks.

INTRODUCTION

This Topic on organic chemistry continues the interpretation of reactions in terms of nucleophilic or electrophilic reagents but also makes extensive use of concepts such as pK_a and hydrogen bonding. The reading on amino acids, proteins and enzymes is relatively extensive because of its relevance to the biological sciences.

A knowledge of the subject matter of this topic is assumed in both the Special Studies *Biochemistry* and *Food science*. It is desirable, therefore, that this topic is covered before either of these Special Studies is attempted.

13.1
CARBOXYLIC ACIDS

Objectives

1 To investigate the reactions of the carboxylic acids.
2 To interpret the reactions in terms of acidic properties and nucleophilic reactions.
3 To learn about some naturally occurring carboxylic acids.

Timing

Less than a week.

Suggested treatment

For this treatment, overhead projection transparency number 110 will be useful. This is a short section, introduced in the *Students' book* by a discussion of possible reaction mechanisms, and the infra-red absorption spectrum of ethanoic acid.

The theory of the nucleophilic formation of esters is potentially difficult. Students are expected to know and be able to interpret the evidence based on the use of the isotope ^{18}O but are not expected to learn the full theory of ester formation.

EXPERIMENT 13.1
An investigation of the reactions of ethanoic acid

Each student or pair of students will need:
Test-tubes and rack
Small beaker
Dropping pipette
Ethanoic acid, pure (glacial), 2 cm^3
0.1M ethanoic acid, 10 cm^3
Full-range Indicator
Phosphorus pentachloride
0.1M sodium hydroxide, 15 cm^3
1M sodium carbonate
Approximately 1M sodium ethanoate solution

Optional:
Apparatus for reflux and distillation, consisting of a 50 cm^3 pear-shaped flask, still head fitted with thermometer, 0–250 °C, and Liebig condenser.
Benzoic acid, 8 g
Concentrated sulphuric acid, 2 cm^3
Methanol, 15 cm^3
0.5M sodium carbonate, 15 cm^3
Sodium sulphate, anhydrous
1,1,1-trichloroethane, 15 cm^3

Procedure

Full instructions are given in the *Students' book*. The laboratory preparation of methyl benzoate is an optional experiment. The following notes may be helpful.

1 Solubility and pH. The first pK_a value of carbonic acid is 6.4 and of phenol is 9.9, so the carboxylic acids release CO_2 from carbonates but phenol does not.

2 Formation of salts. A gradual change of pH should be observed, characteristic of a weak acid forming a buffer solution with its salt. Sodium ethanoate is alkaline because of salt hydrolysis.

3 Formation of the acid chloride. This reaction is discussed in part three of the section on the reactions of carboxylic acids in the *Students' book*.

4 Laboratory preparation of the ester methyl benzoate. This is an optional experiment, which need only be done if it is not intended to carry out Experiment 17.5 'A problem in synthesis'. The methyl benzoate prepared by students can be used for the nitration experiment in Topic 9 (Experiment 9.5b of *Students' book I*).

Reactions of carboxylic acids

A review of the reactions of the carboxylic acids is given in the *Students' book*. In several of the more complicated cases full balanced equations are not used; a simplified version is all that is required. Reduction using lithium aluminium hydride, for example, is shown thus:

$$\text{C}_6\text{H}_5\text{-CO}_2\text{H} \xrightarrow{\text{LiAlH}_4} \text{C}_6\text{H}_5\text{-CH}_2\text{OH} + \text{H}_2$$

For the teacher's information the full balanced equation is:

$$4\,\text{C}_6\text{H}_5\text{-CO}_2\text{H} + 3\text{LiAlH}_4 + 12\text{HCl} \longrightarrow 4\,\text{C}_6\text{H}_5\text{-CH}_2\text{OH} + 4\text{H}_2 + 3\text{LiCl} + 3\text{AlCl}_3 + 4\text{H}_2\text{O}$$

An opportunity should be taken to test the students' knowledge of these reactions at a suitable point.

Background reading

This section in the *Students' book* ends with some background reading entitled 'Naturally occurring carboxylic acids'.

Supporting homework

Learning the reactions of ethanoic acid.
Answering questions from the end of the Topic in the *Students' book*.
Reading the Background reading 'Naturally occurring carboxylic acids'.

Summary

At the end of this section students should:
    1 know the reactions of ethanoic acid that they have encountered;
    2 be aware of what lipids are and their importance in nature and in industry.

13.2
CARBOXYLIC ACID DERIVATIVES
Objectives

1 To investigate the reactions of carboxylic acid derivatives.
2 To learn about liquid crystals and LCDs.

Timing

Less than a week.

Suggested treatment

Students often have difficulty with the formulae of the acid derivatives. The building of ball-and-spoke models to help with the interpretation of the reactions is recommended. One preparation of cholesteryl benzoate provides enough material for a whole class to examine its liquid crystal properties.

EXPERIMENT 13.2
An investigation of some reactions of carboxylic acid derivatives

The teacher will need:
3 small beakers
Dropping pipette
Ethanoyl chloride
Ethanol
Sodium carbonate solution
0.880 ammonia solution

Each student or pair of students will need:
Apparatus for reflux consisting of a 50 cm^3 pear-shaped flask and Liebig condenser
10 cm^3 measuring cylinder
Ethanamide
2M hydrochloric acid
Methyl benzoate, 2 cm^3
2M sodium hydroxide, 35 cm^3

Optional:
Access to a fume cupboard
Apparatus for suction filtration: 100 cm^3 Buchner flask and filter funnel.
Bunsen burner and tongs
50 cm^3 conical flask
Ice bath
Microscope slide
Piece of polaroid (mentioned in background reading)
Steam bath
Benzoyl chloride, 0.4 cm^3
Cholesterol, 1 g
Ethyl ethanoate, 20 cm^3
Methanol, 20 cm^3
Pyridine, 3 cm^3

Procedure

Full instructions are given in the *Students' book*. THE EXPERIMENTS WITH ETHANOYL CHLORIDE SHOULD BE DEMONSTRATED; the teacher should use ethanoyl chloride with caution. It is volatile and forms pungent fumes in moist air, and will react violently with anything containing water, and with alcohols. The optional preparation of cholesteryl benzoate is best carried out in a fume cupboard. The following notes may be helpful.

1 Acid chlorides. The reactions are violent and HCl fumes are evolved.

2 Esters. Students should set up this experiment before the teacher demonstration of experiment 1.

3 Acid amides. Amides are hydrolysed by both acid and alkali catalysts.

4 Laboratory preparation of cholesteryl benzoate, 'liquid crystals'. This is an optional experiment but the properties of the product should be demonstrated. The safety precautions concerning the use of pyridine (harmful vapour; avoid contact with skin and eyes) and benzoyl chloride (lachrymatory) mentioned in the *Students' book* should be rigidly adhered to.

Reactions of carboxylic acid derivatives

The *Students' book* contains a survey of the principal reactions of the carboxylic acid derivatives. An opportunity should be taken to test the students' knowledge of these reactions at a suitable point.

Supporting material

Ball-and-spoke molecular model kits.

Supporting homework

Learning the reactions of carboxylic acid derivatives.
Preparing an account of suitable reactions for preparation of carboxylic acid derivatives.
Answering questions from the end of this Topic in the *Students' book*.
Reading the Background reading 'Liquid crystal displays'.

Summary

At the end of this section students should:

1 know the reactions of the carboxylic acid derivatives they have encountered;

2 be aware of what is meant by a liquid crystal and how liquid crystals are used in LCDs (liquid crystal displays).

13.3
AMINES

Objectives
To investigate the reactions of amines.

Timing
Less than a week.

Suggested treatment
For this treatment overhead projection transparency number 111 will be useful.

The diazotization reaction is studied in Topic 17 as part of a section on dyestuffs but it could be taken now if wished.

The difference in structure between primary, secondary, and tertiary amines and the parallel alcohols (Topic 11) and halogenoalkanes (Topic 9) should be pointed out by the use of models as well as the links with nitriles (reduction reaction) and amides (ethanoylation reaction).

EXPERIMENT 13.3
An investigation of the reactions of amines

Each student or pair of students will need:
Test-tubes and rack
Dropping pipette
2M ammonia
Butylamine
0.1M copper(II) sulphate solution
Ethanoyl chloride
Full-range Indicator
Phenylamine
2M hydrochloric acid

Optional:
1 Apparatus for reflux, consisting of 50 cm³ pear-shaped flask and Liebig condenser
   Access to darkened room
   Apparatus for suction filtration (100 cm³ Buchner flask and funnel)
   2 watchglasses
   2-aminobenzoic acid, 3.5 g
   Ethanoic anhydride, 10 cm³
   Methanol
2 Apparatus as in figure 13.1
   50% aqueous ethanol
   Decanedioyl dichloride, 0.5 cm³
   Hexane-1,6-diamine, 0.7 g
   Sodium carbonate, anhydrous, 2 g
   Litmus
   1,1,2,2-tetrachloroethane (or permitted high density solvent), 15 cm³

Figure 13.1
The 'nylon rope trick'.
Photograph, University of Bristol, Faculty of Arts Photographic Unit.

Procedure

Full instructions are given in the *Students' book.* Warn students that phenylamine is harmful by skin absorption. If the optional laboratory preparations are attempted, both can readily be carried out by different students in the class. The following notes may be helpful.

1 Solubility and pH. If a pH meter is available, it is worth while comparing the pH of 0.1M ammonia and 0.1M butylamine. Students tend to assume that inorganic compounds are stronger acids or bases than organic compounds.

2 Formation of salts. If students write out the equations, they should see that the pattern of the reactions is the same as for the familiar ammonium salts.

3 Reaction with metal ions. The amines are reacting as ligands and forming complex ions.

4 Ethanoylation. Students must be reminded of the violence of reactions involving ethanoyl chloride. If there is the slightest doubt about the students' ability to carry out this reaction safely it should be demonstrated. If students lack experience in the determination of melting point, they should dry their product and determine its melting point.

5 Laboratory preparations. These are optional experiments.

Reactions of amines

The *Students' book* contains a survey of the principal reactions of amines. An opportunity should be taken to test the students' knowledge of these reactions at a suitable point.

Supporting material

Ball-and-spoke molecular models.

Supporting homework

Learning the reactions of the amines.
Answering questions from the end of this Topic in the *Students' book*.

Summary

At the end of this section students should:
1 know the reactions of ammonia, butylamine, and phenylamine they have encountered;
2 be aware of what is meant by triboluminescence;
3 know how to prepare nylon (or this may be delayed until Topic 17).

13.4
AMINO ACIDS AND PROTEINS
Objectives

1 To learn about the reactions and natural occurrence of amino acids.
2 To gain skill in the technique of paper chromatography.
3 To become aware of the structural complexity of proteins and the α-helix.

Timing

About a week.

Suggested treatment

For this treatment overhead projection transparency number 112 will be useful.

The section begins with a consideration of naturally occurring amino acids. The following points should be noted.

1 The non-systematic names generally used by biochemists are used in the *Students' book*.

2 The attention of students should be directed to the existence of a chiral centre in the molecules of all of these compounds except for glycine; the opportunity can be taken to revise this aspect of isomerism, which was first mentioned in Topic 11.

3 Amino acid 'residues' can be linked together through the —CO—NH— group. In this context this group is known as the *peptide* group. Two amino acid residues form a *dipeptide*; three a *tripeptide*; a number of amino acid residues form a *polypeptide*.

Protein molecules have very large numbers of amino acid residues linked together through peptide groups.

4 All proteins are made up from about two dozen amino acids. Twenty of them are listed in the *Students' book*.

The main objective of a discussion of proteins should be to establish their importance rather than to learn facts about specific proteins. Additional information is given in the Special Studies *Food science* and *Biochemistry*, and biology students should be able to make a valuable contribution to the discussion.

The teaching of this section can be merged with the teaching of the Special Studies *Food science* or *Biochemistry*, if either of these studies is being taken by the students.

EXPERIMENT 13.4a
An investigation of protein materials

Each student or pair of students will need:

Access to fume cupboard
Access to oven or Bunsen burner
Simple polarimeter (optional)
Test-tubes and rack
Dropping pipette
Chromatography paper
0.1M copper (II) sulphate solution
Full-range Indicator

0.01M hydrochloric acid
0.01M L-glutamic acid
0.01M glycine
2M sodium hydroxide
0.01M sodium hydroxide
0.02M ninhydrin solution in propanone
Protein materials

Procedure

Full instructions are given in the *Students' book*. Because Experiment 13.4b 'The chromatographic separation of amino acids' is lengthy, students could start on it and then carry out Experiment 13.4a while their chromatograms develop.

The following notes may be helpful.

1 *Acidity and basicity.* Glycine and L-glutamic acid are used for these experiments as they are the cheapest amino acids to purchase. The pH of other amino acids could be demonstrated to students. Students need not be introduced to the term 'zwitterion'.

2 *Biuret test.* This test depends on the formation of a complex ion between copper(II) ions and peptide groups.

3 *Ninhydrin test.* Students should be cautioned against getting ninhydrin spray on their hands. The reactions are complex and students are not expected to know them; for the teacher's information, they can be written as follows.

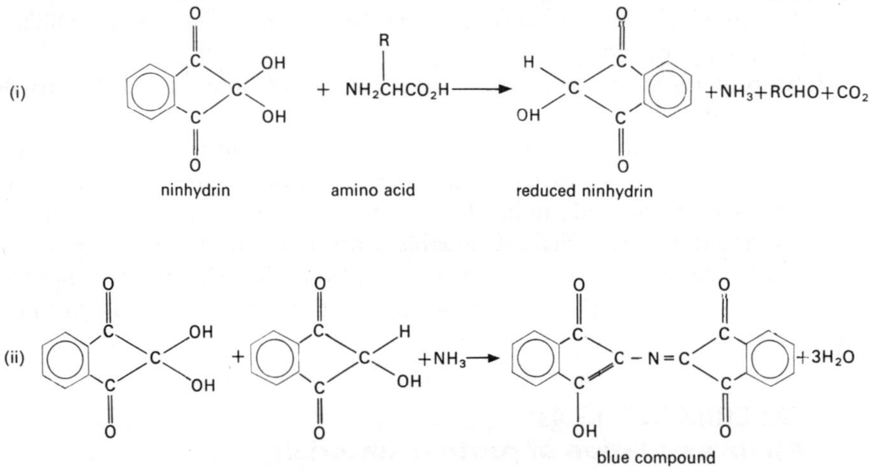

(i) ninhydrin + amino acid (NH_2CHRCO_2H) $\longrightarrow$ reduced ninhydrin + $NH_3 + RCHO + CO_2$

(ii) ninhydrin + reduced ninhydrin + $NH_3 \longrightarrow$ blue compound + $3H_2O$

Figure 13.2
The development of ninhydrin blue.

4 *Chirality.* Students should appreciate by the inspection of models that glycine is achiral while the other amino acids are chiral. A stock solution of sodium glutamate can be made sufficiently concentrated to display slight rotation in a simple polarimeter. If students use the stock solution to make their measurements, not much is needed and the experiment is quickly performed.

EXPERIMENT 13.4b
The chromatographic separation of amino acids

Each student or pair of students will need:
0.01% amino acid solutions (see procedure)

Apparatus for paper chromatography:
1 dm^3 beaker and cover (cling film)
Capillary melting point tubes
Chromatography paper (No. 1 paper, 12.5 cm reel)
25 cm^3 measuring cylinder
5 cm^3 measuring cylinder
Paper clip

Access to:
0.880 ammonia in crystallizing dish (*in fume cupboard*)
Butan-1-ol, 12 cm^3
Ethanoic acid, pure (glacial), 3 cm^3
Oven at 110 °C
Spray bottle containing methanol (19 cm^3), 1M aqueous copper(II) nitrate (1 cm^3), and 2M nitric acid (1 drop)
Spray bottle containing 0.02M ninhydrin in propanone (store in refrigerator)

Procedure

This experiment is lengthy and if convenient, students could start it and carry out the preceding Experiment 13.4a while their chromatograms develop. Full instructions are given in the *Students' book*. The following notes may be helpful.

Reference solutions of 0.01% amino acids should be made up in a mixture of water (9 parts) and propan-2-ol (1 part). Reference sets of amino acids can be purchased or about five of the cheaper amino acids could be selected to cover the range of polar (asparagine, tyrosine), non-polar (alanine, glycine, leucine, valine), acidic (aspartic acid, glutamic acid), and basic (arginine, lysine). To reduce waste 25 cm lengths of chromatography paper can be cut from the reel in advance. Spraying of chromatography papers should be carried out in a fume cupboard with the paper hung up and *not* held in an unprotected hand. Ninhydrin aerosol sprays suitable for use in this experiment can be obtained from BDH Chemicals Ltd.

This section in the *Students' book* ends with a piece of Background reading entitled 'The chemical and structural investigation of proteins'. This is quite a difficult subject and may require some explanation in the classroom.

Supporting material

Ball-and-spring molecular model kit.
Unilever film 'The structure of protein'. 16 mm, colour, sound, 17 minutes.
Available for hire from: The Scottish Central Film Library, Dowanhill, 74 Victoria Crescent Road, Glasgow, G12 9JN.
Unilever Educational Booklet, Advanced Series No. 3 *The chemistry of proteins.* (1981) Available from: Unilever Education Section, P.O. Box 68, Unilever House, London EC4P 4BQ.

Supporting homework

Collecting protein material for laboratory testing.
Learning the properties of amino acids.
Reading the Background reading 'The chemical and structural investigation of proteins'.

Summary

At the end of this section students should:
    1 know what is meant by chirality, peptide group, and polypeptide;
    2 be acquainted with the 20 most important amino acids in nature;
    3 have experience of the technique of paper chromatography;
    4 know in general terms how proteins can be analysed into different arrangements of amino acids.

13.5
ENZYMES
Objectives

1 To learn about the catalytic properties of enzymes.
2 To become aware of the biochemical, industrial, and medical importance of enzymes.

Timing
Three periods.

Suggested treatment

Students study an enzyme catalysed reaction, noting the specific nature of the enzyme. In further discussion the sensitivity of enzyme activity to temperature and pH of solution is introduced.

These characteristics of enzyme catalysis can be compared with the characteristic features of inorganic catalysts when these are studied later in the course (Topics 14 and 16).

EXPERIMENT 13.5
An investigation of the enzyme-catalysed hydrolysis of urea

Each student or pair of students will need:
Test-tubes and rack
Dropping tube
Bunsen burner
Stopclock
Full-range Indicator
0.25M ethanamide, 5 cm^3
0.01M hydrochloric acid, 10 cm^3
0.25M methylurea, 5 cm^3
0.25M urea, 5 cm^3
1 % urease active meal solution (a cloudy suspension), 5 cm^3

Procedure

Full instructions are given in the *Students' book*. Students should find that urease will only hydrolyse urea. The boiling of enzymes causes a loss of structure due to hydrogen bonds breaking and is known as 'denaturing'.

The *Students' book* lists a number of properties of enzymes and the study of enzymes is taken further in the Special Study *Biochemistry*. The *Students' book*

56 Topic 13 Carbon compounds with acidic and basic properties

at this point contains a description of the activities of biochemists, and what distinguishes them from organic chemists; and two pieces of Background reading 'The industrial uses of enzymes' and 'Enzymes in medicine'.

Supporting homework

Reading the Background reading 'The industrial uses of enzymes' and 'Enzymes in medicine'.

Summary

At the end of this section students should:
1 know some of the properties of enzymes;
2 be aware of the importance of enzymes to biological processes;
3 be aware of some applications of enzymes in industry and in medicine.

13.6
SURVEY OF REACTIONS IN TOPIC 13

As in Topics 9 and 11, the chart provided in the *Students' book* is intended as a base from which students can develop their own more elaborate charts. It is suggested that a set of charts each with different information such as formulae, reagents or type of reaction, is more valuable for revision purposes than a single chart containing all the information.

Timing

One period plus homework.

ANSWERS TO PROBLEMS IN THE *STUDENTS' BOOK*

(A suggested mark allocation is given in brackets after each answer.)

Carboxylic acids

1 Butane   Van der Waals forces only.   (2)
   Propanal   Dipole-dipole interactions.   (2)
   Propan-1-ol   Hydrogen bonding.   (2)
   Ethanoic acid   Hydrogen bonding and 2 polar atoms in same
                   molecule   (2)
   Van der Waals forces are of similar magnitude in each compound:
   hydrogen bond interactions are larger than dipole-dipole attractions.   (2)
   Total 10 marks

2a B; A; D; C. (2)
b A; B; C; D. (2)
c A; E; D; C; B. (2)
Discuss acid strength in terms of stabilization of delocalized charge on anions by electron-withdrawing groups. Give some credit for discussion of the effect of electron-withdrawing groups on the ease of ionization of the —O—H bond in the acid. (4)

Total 10 marks

3a 1-iodo-4-methylpentane. (1)
b Reflux with an aqueous solution of sodium hydroxide. Nucleophilic substitution reaction. (2)
c Hot, concentrated, alcoholic solution of sodium hydroxide. Elimination reaction. (2)

d

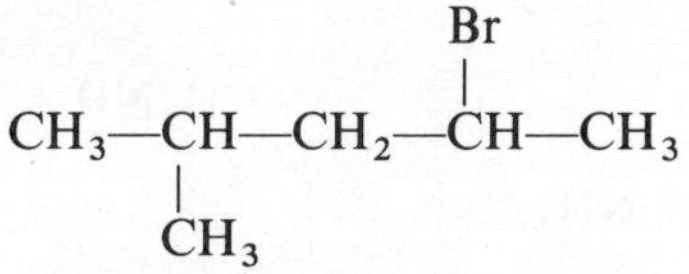

$$CH_3-CH-CH_2-\underset{}{\overset{Br}{CH}}-CH_3$$
$$\underset{CH_3}{|}$$

2-bromo-4-methylpentane. (2)

Electrophilic addition reaction. (1)

e $CH_3-\underset{\underset{CH_3}{|}}{CH}-CH_2-CH_2-CHO$ (1)

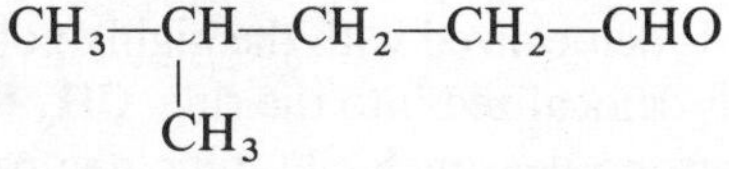

f PCl_3 or PCl_5 (1)

g

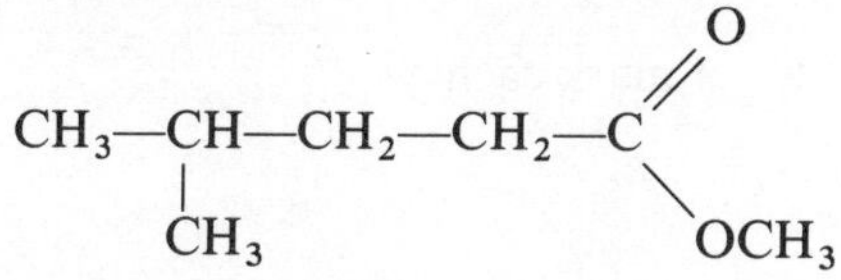

$CH_3-\underset{\underset{CH_3}{|}}{CH}-CH_2-CH_2-C\begin{smallmatrix}\diagup\mathrm{O}\\ \diagdown\mathrm{OCH_3}\end{smallmatrix}$ (1)

h $CH_3-\underset{\underset{CH_3}{|}}{CH}-CH_2-CH_2-CH_2OH$ (1)

i Substance C. (1)

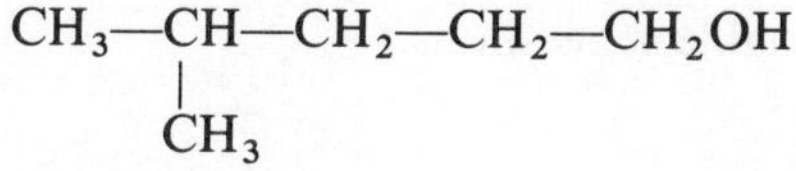

j $CH_3-\underset{\underset{CH_3}{|}}{CH}-CH_2-CH_2-COCl$ (1)

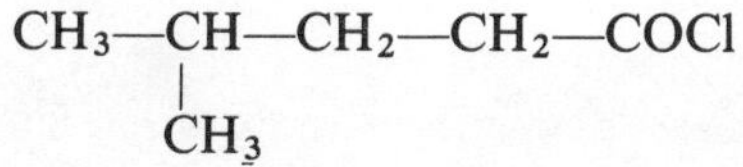

Total 14 marks

4 Any appropriate series of reactions. **Total 10 marks**

5 Free response question: mark by impression. **Total 20 marks**

Amines

6 CH_3—CH_2—CH_2—CH_3; CH_3—CH_2—CH_2—NH_2;
CH_3—CH_2—CH_2OH; CH_3—$CONH_2$. (2)
Van der Waals forces are similar in all substances. Hydrogen bonding interactions in all cases except for butane, but —OH with 2 lone-pairs and greater polarity has stronger interactions than —NH_2. Amide has two electronegative atoms for H-bonding or dipole-dipole attractions. (4)

Total 6 marks

7 [structures: benzamide (C6H5—C(=O)—NH2); phenylamine (C6H5—NH2); NH_3; benzylamine (C6H5—CH2NH2)] (2)

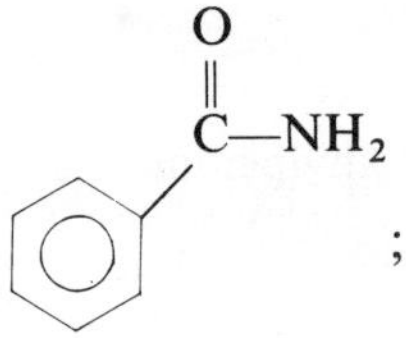

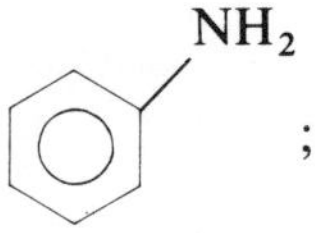

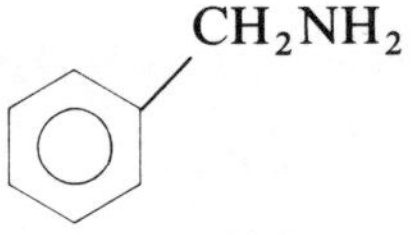

Discuss in terms of availability of electron lone-pair for donation to protons. In the amide the lone-pair is delocalized with the highly polar carbonyl group, in phenylamine it is delocalized into the ring. CH_3— or —CH_2— groups are weakly electron-releasing facilitating lone-pair donation in the strongest base. (4)

Total 6 marks

8a i A; ii B; iii D; iv F; v E. 1 mark each (5)
 b A, B, D. (2)
 c A. (1)
 d C and E. (2)
 e E and F. (2)
 f A, B, D. (E and F are also reasonable suggestions.) (2)
 g A and B. (2)
 h D. (1)
 i E. (1)

Total 18 marks

9 Alternative answers are possible for A, B, and C in (i).

i A $CH_3-CH_2-CH_2-CH-CH_3$ or $CH_3-CH_2-CH-CH_2NH_2$
               $\quad\quad\quad\quad\quad\quad\quad |$ $\quad\quad\quad\quad |$
               $\quad\quad\quad\quad\quad\quad\quad NH_2$ $\quad\quad\quad\quad CH_3$ (2)

   B $CH_3-CH_2-CH_2-CH-CH_3$ or $CH_3-CH_2-CH-CH_2NH_3^+Cl^-$
               $\quad\quad\quad\quad\quad\quad\quad |$ $\quad\quad\quad\quad |$
               $\quad\quad\quad\quad\quad\quad\quad NH_3^+Cl^-$ $\quad\quad\quad\quad CH_3$ (2)

   C $CH_3-CH_2-CH_2-CH-CH_3$ or $CH_3-CH_2-CH-CH_2NHCOCH_3$
               $\quad\quad\quad\quad\quad\quad\quad |$ $\quad\quad\quad\quad |$
               $\quad\quad\quad\quad\quad\quad\quad NHCOCH_3$ $\quad\quad\quad\quad CH_3$ (2)

ii $CH_3-CH_2-CH-CH_2-CH_3$ and
             $\quad\quad\quad\quad |$
             $\quad\quad\quad\quad NH_2$

   $CH_3-CH_2-CH_2-CH_2-CH_2NH_2$ (2)

iii a $CH_3-CH_2-CH_2-CH_2-CH_2OH \xrightarrow[\text{KBr + conc. H}_2\text{SO}_4]{\text{HBr from}}$

   $CH_3-CH_2-CH_2-CH_2-CH_2Br \xrightarrow[\text{alcoholic solution of ammonia}]{\text{Heat under pressure with}}$

   $CH_3-CH_2-CH_2-CH_2-CH_2NH_2$ (4)

   b $CH_3-CH_2-CH_2-CH_2OH \xrightarrow{\text{HBr}}$

   $CH_3-CH_2-CH_2-CH_2Br \xrightarrow[\text{solution of NaCN}]{\text{Reflux with alcoholic}}$

   $CH_3-CH_2-CH_2-CH_2-CN \xrightarrow[\text{H}_2/\text{Ni catalyst}]{\text{Reduce}}$

   $CH_3-CH_2-CH_2-CH_2-CH_2-NH_2$ (4)

Total 16 marks

10 Free response question: mark by impression.   **Total 10 marks**

Acid derivatives

11a D. (1)
 b C. (1)
 c B. (1)

 d sodium benzoate: benzene ring–C(=O)–O⁻Na⁺ (1)

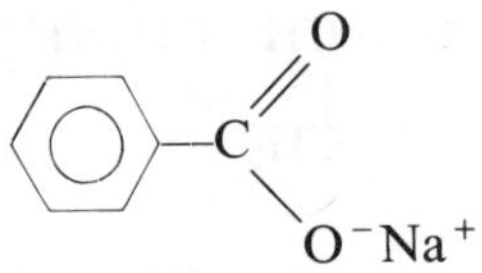

 e PCl$_3$ or PCl$_5$ (1)
 f CH$_3$OH (1)
 g NH$_3$ (1)
 h CH$_3$OH and sodium benzoate (benzene ring–C(=O)–O⁻Na⁺) (2)

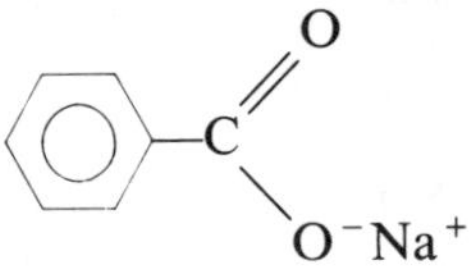

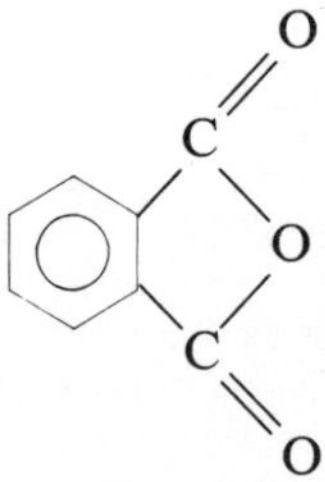

 i B and E (2)

 j benzene-1,2-dicarboxylic anhydride (phthalic anhydride) and benzonitrile (C$_6$H$_5$–C≡N) (2)

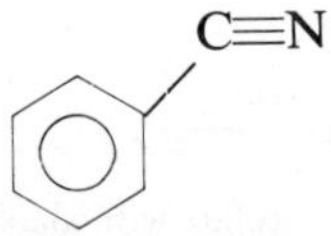

Total 13 marks

12a To avoid reaction with moisture in the air. (1)
 b To allow the dehydration reaction to go to completion. (1)
 c Release excess pressure from any CO$_2$ which is formed. (1)
 d To release ethanonitrile from aqueous solution ('salting out'). (1)
 e Only small quantities of a weak acid require neutralization and a strong base might catalyse the hydrolysis of ethanonitrile. (2)
 f Hydrolysis of ethanamide. (1)
 g Wash with water, dry product over suitable drying agent and distil. (3)

Total 10 marks

13a $CH_3CO_2CH_3$ (2)
 b $CH_3CO_2C_2H_5$ (2)
 c CH_3CONH_2 (2)
 d $CH_3C\equiv N$ (2)
 e $CH_3CH_2NH_2$ (2)
 Total 10 marks

14 Any suitable reactions. (2 marks each) (10)

Amino acids and proteins

15a i Amino group, carboxyl group and both attached to C—H. (3)
 ii Two carboxyl groups, four different groups attached to central carbon atom and longer carbon chain. (3)
 b i Both are amino acids, possible salt formation with $-NH_2$ and $-CO_2H$ groups. (3)
 Other properties of these groups. max. (3)
 ii Aspartic acid could exist as two optical enantiomers; glycine could not. The presence of two carboxyl groups in aspartic acid would make it more acidic than glycine. (2)
 Total 14 marks

16 Charges on the molecule vary with pH.

$H_3\overset{+}{N}-CH_2-CO_2^-$ $H_3\overset{+}{N}-CH_2-CO_2H$ $H_2N-CH_2-CO_2^-$ (6)
pH 7   pH 2   pH 12

17 **a, b** and **c** No difference. (3)
 d One will rotate the plane of plane polarized light in a clockwise sense and the other anti-clockwise by the same amount. (2)
 e Naturally occurring amino acids are all of the L form which, strictly speaking, could be either ($+$) or ($-$) in natural protein. It is reasonable for answers to suppose that either ($+$) or ($-$) would occur alone in any particular protein. (2)
 Total 7 marks

18a C, O, and N. (1)
 b C—H bond. (2)

 c (4)

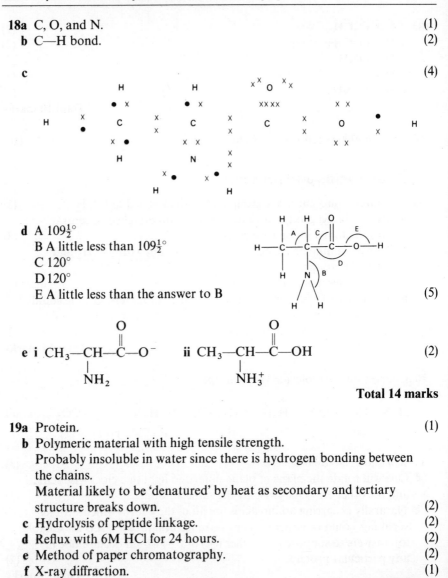

 d A $109\tfrac{1}{2}°$
   B A little less than $109\tfrac{1}{2}°$
   C $120°$
   D $120°$
   E A little less than the answer to B (5)

 e i CH_3—CH(NH$_2$)—C(=O)—O$^-$ ii CH_3—CH(NH$_3^+$)—C(=O)—OH (2)

Total 14 marks

19a Protein. (1)
 b Polymeric material with high tensile strength.
   Probably insoluble in water since there is hydrogen bonding between the chains.
   Material likely to be 'denatured' by heat as secondary and tertiary structure breaks down. (2)
 c Hydrolysis of peptide linkage. (2)
 d Reflux with 6M HCl for 24 hours. (2)
 e Method of paper chromatography. (2)
 f X-ray diffraction. (1)

Total 10 marks

TOPIC 14
Reaction rates – an introduction to chemical kinetics

OBJECTIVES

1 To study the factors which influence the rate of reactions.
2 To show how kinetic studies provide evidence in support of reaction mechanisms.
3 To introduce a simple treatment of the collision theory of reaction kinetics.
4 To gather information on catalysts, and to introduce some theoretical understanding of catalysis.

CONTENT

14.1 Introduction. Factors which may influence rates; why we study rates of reaction.
14.2 The effect of reactant concentration on the rate of a reaction. Rate equations; the order of a reaction; rate constants; problems involved in measuring rates; experimental investigation of the effect of concentration; review of methods available for following the course of a reaction; kinetics and reaction mechanisms.
14.3 The effect of temperature on the rate of a reaction. Experimental investigation; the collision theory of reaction kinetics; activation energies; the Arrhenius equation.
14.4 Catalysis. Recall of knowledge of catalysts; catalysts provide a reaction pathway of lower activation energy; experimental investigation of a system involving autocatalysis.

TIMING

Three weeks.

SUPPORTING MATERIAL

The Educational Techniques Subject Group of the Royal Society of Chemistry has produced a resource book on reaction kinetics, which includes details of experiments, kinetics data, and a survey of school practice in the United Kingdom. The publication is: *Reaction kinetics: a resource for teachers*, by B. E. Dawson, C. L. Mason, and P. Mason, Royal Society of Chemistry ETSG, 1981.

The book is available from the Scientific Affairs Officer, The Royal Society of Chemistry, Burlington House, Piccadilly, London W1V 0BN.

INTRODUCTION

The students will, no doubt, have some previous qualitative experience of rates of reaction, possibly backed up by some experimental measurements. Even if formal experience is lacking it should be possible with careful questioning to establish sufficient background; but if it is thought desirable the subject could be introduced by doing some experiments from Revised Nuffield Chemistry, *Teachers' guide II*, Topic A18.

14.1
INTRODUCTION
Objectives

1 To consider the factors which might influence reaction rates.
2 To make students aware of the reasons for studying reaction rates.

Timing

If the students have some previous experience of rate studies, a single period should be sufficient; in some cases the students could read the section for homework and move on to the next section immediately. If the students have no previous experience, of course, more time will be needed.

Suggested treatment

With the aid of some directed questioning from the teacher, the students should be able to recall, or suggest, factors which are likely to influence the rates of reactions. A list of such factors is given in the *Students' book*.

The students are about to embark on a study of a fairly detailed subject which will probably appear to many to be more physical and mathematical than chemical. It is important at the outset to explain why they will be going to the trouble of *measuring* reaction rates rather than being content with a qualitative knowledge. There is no need to labour the point – what is given in the *Students' book* should be enough to establish a strategy for the topic.

Summary

After this section, the students should be aware of the factors likely to influence rates of reactions, and understand the reasons for studying them.

14.2
THE EFFECT OF REACTANT CONCENTRATION ON THE RATE OF A REACTION

Objectives

1 To study rate equations.
2 To consider the problems involved in measuring rates.
3 To obtain, and use, experimental evidence to establish orders of reaction.
4 To consider various experimental techniques for the measurement of rates.
5 To show how a knowledge of orders of reactions can provide evidence in support of reaction mechanisms.

Timing

The description and discussion of the rate equation will take two periods. Experiment 14.2a and its analysis will take a double period. The survey of methods of following a reaction could be set for private reading. Experiment 14.2b will require a double period plus time for graph-drawing, etc. The mechanistic discussions may well take three double periods, four if Experiment 14.2c is also done. In all, about two weeks will be needed.

Suggested treatment

There is much scope for individual variation of treatment here, particularly in the experimental work.

The general idea of a rate equation is first introduced, and each term in it is discussed. The rate equation is an equation showing how the rate of change of concentration of a substance A, symbol r_A, depends upon the concentrations of the various substances involved in the reaction.

When discussing this equation, it is particularly important to see that the students gain a clear understanding of the definitions of the various terms. Although the term *rate of reaction* is perfectly acceptable in a general sense, and when referring to purely qualitative observations, teachers should note that it should *not* be used in a quantitative sense, at any rate in schools, so that any possible ambiguity is avoided (see *Chemical nomenclature, symbols, and terminology*, Association for Science Education, 1984). The rate equation does not describe the 'rate of the reaction' but the rate of change of concentration of one named substance taking part in the reaction. It is clearly important to state to which substance the equation refers.

The *Students' book* gives a typical rate equation

$$r_A = k[A]^a[B]^b[C]^c$$

Each term in this rate equation is then defined. When discussing the definitions, bear in mind that students do not always realize unaided that one of the implications of a rate equation of the form

$$r_A = k[A][B]$$

is that increasing the concentrations of both A and B by a factor of two increases r_A by a factor of four.

The idea that the rate constant k is a measure of the rate of change of concentration of substance A, r_A, at unit concentration of each of the substances appearing in the rate equation needs a little stress at this point, because it is implied in the discussion of the results of Experiment 14.3.

EXPERIMENT 14.2a
The kinetics of the reaction between calcium carbonate and hydrochloric acid

This is a simple experiment which gives rapid results. Two versions are offered. Method 1, using a direct-reading balance, could be done as a teacher demonstration with students assisting; Method 2 is suitable for students themselves.

Method 1

For this method, the teacher will need:
Top-loading, direct-reading balance Marble, lumps, 10 g
Measuring cylinder, 50 cm^3 1M hydrochloric acid, 20 cm^3
Cotton wool Watch or clock reading seconds
Conical flask, 100 cm^3

Method 2

For this method, each student or pair of students will need:
Test-tube with side-arm, 150 × 25 mm Marble, lumps, 10 g
Glass gas syringe, 100 cm^3 1M hydrochloric acid, 10 cm^3
Rubber stopper to fit test-tube Watch or clock reading seconds
Rubber connection tubing

Procedure

Full details are given in the *Students' book*. It is important that the acid should be saturated with carbon dioxide before readings are begun. In Method 2, the maximum volume of carbon dioxide obtainable from 10 cm^3 of 1M hydrochloric acid is about 120 cm^3, so the timing must not be started too early or the syringe will be overfilled. Sample results are shown in the graphs in figure 14.1.

14.2 The effect of reactant concentration on the rate of a reaction

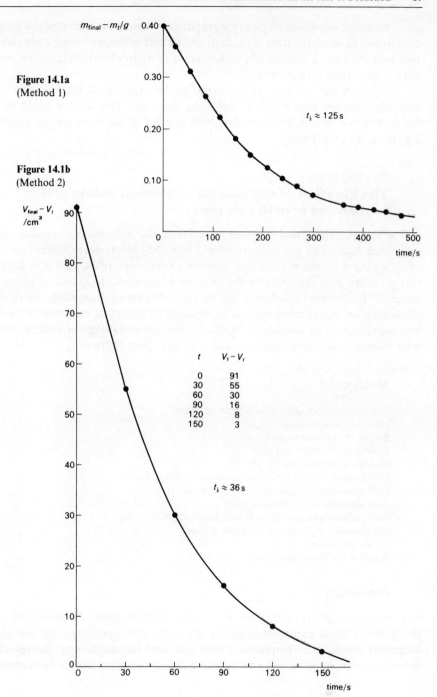

Figure 14.1a (Method 1)

Figure 14.1b (Method 2)

t	$V_f - V_t$
0	91
30	55
60	30
90	16
120	8
150	3

Students are asked to plot this graph, and compare it with types of graphs that would be expected from zero, first, and second order reactions. They should find that the rate of change of concentration of hydrochloric acid is first order with respect to hydrochloric acid.

After the instructions and discussion of this experiment, the *Students' book* contains a survey of methods of following reactions. This section could be set for homework, or discussed in class with the aid of the examples in question 1 at the end of the Topic.

EXPERIMENT 14.2b
The kinetics of the reaction between iodine and propanone in acid solution

For a second experiment on rates of reaction, the iodination of propanone is suggested. Again, two methods are offered. Method 1 illustrates a different method of following a reaction from that used in Experiment 14.2a, and it is used at this point because the order of the reaction with respect to iodine turns out to be zero. This point is followed up in the subsequent discussion. Method 2 illustrates an 'initial rate' method of arriving at orders of reaction. Although less sophisticated in technique, Method 2 has the advantage of finding orders with respect to all three reactants, and is quicker than Method 1.

Method 1

Each student or pair of students will need:
4 or more titration flasks
Burette, 50 cm^3, and burette stand
Pipette, 10 cm^3, and safety filler
Measuring cylinder, 50 cm^3
0.02M iodine in potassium iodide solution, labelled A, 50 cm^3
1.0M propanone solution (in water), labelled B, 25 cm^3
1.0M sulphuric acid, labelled C, 25 cm^3
0.5M sodium hydrogen carbonate, labelled D, 150 cm^3
0.01M sodium thiosulphate, labelled E, 150 cm^3
1% starch solution, 10 cm^3
Watch or clock reading seconds

Procedure

Full details are given in the *Students' book*. It will probably be helpful to give the students some explanation before they begin. In particular, the sampling technique needs to be emphasized, and the need for 'quenching' the reaction before titration, in this case by neutralizing the acid catalyst, should be discussed.

Method 2

Each student or pair of students will need:
Conical flask, 100 cm^3
Test-tube

Access to communal burettes containing:
2M hydrochloric acid, 100 cm^3 per student
2M propanone, 40 cm^3 per student
0.01M iodine in potassium iodide solution, 20 cm^3 per student
Watch or clock reading seconds

Quantities of solutions given are the total quantities likely to be required by each student or pair of students.

Procedure

Details of the procedure to be followed are given in the *Students' book*. It is important that the volumes are measured as accurately as possible, because of the small quantities that are used. This method is particularly impressive if all four runs are done simultaneously, quite possible if students work in groups.

Sample results

	Run 1	Run 2	Run 3	Run 4
time for colour to disappear/s	115	264	243	58
Rate	$\frac{4}{115}$ = 0.035	$\frac{4}{264}$ = 0.015	$\frac{4}{243}$ = 0.016	$\frac{2}{58}$ = 0.034

Answers to questions

1 Water is added to some of the mixtures to ensure that the total volumes of all four mixtures are the same. In this way the concentration of each reactant is proportional to the volume of solution that is used.

2 If the concentration of acid is halved, the rate of change of concentration of iodine is also halved.

3 The reaction is first order with respect to hydrogen ions.

4 The reaction is first order with respect to propanone, but zero order with respect to iodine.

The mechanistic implications of this experiment are next discussed, and a full account is given in the *Students' book*. The discussion should not be rushed as this subject is often one which students find difficult.

The section ends with a reinforcing discussion of the S_N1 and S_N2 mechanisms for the hydrolysis of halogenoalkanes, described in the *Students' book* as Case A, Case B, and Case C. Overhead projection transparency number

113 will be useful here. Case B uses the same example as Case A; the point of including it is to introduce the initial rate argument, which is then used in Case C, some of the questions at the end of the Topic, and in Experiment 14.3. This treatment is based on work by Dr A. E. Pearson (York University M.Sc. thesis, 1977).

The answers to the questions are:

Case A

1 These results might have been obtained by removing portions of the reaction mixture at measured time intervals, quenching them by rapid cooling in ice, and quickly titrating them with standard acid.

2 It is a second order reaction, as the half-life increases with time.

3 As measured in this experiment, this is an *overall* order, since neither of the two reactants is in large excess.

4 The S_N2 mechanism is in operation in this case. The evidence shows that the reaction is second order; the S_N1 mechanism demands first order kinetics overall.

Case B

1a If the concentration of 1-bromobutane is doubled, the rate of change of concentration of 1-bromobutane also doubles.

b The reaction is first order with respect to 1-bromobutane.

2a If the concentration of hydroxide ion is increased five times, the rate of change of concentration of 1-bromobutane also increases five times.

b The reaction is first order with respect to hydroxide ions.

3 Since the rate of change of concentration of 1-bromobutane depends on the concentration of hydroxide ions, the S_N2 mechanism operates in this case.

Case C

1 The column headed $(V_{final} - V_t)$ represents a measure of the concentration of 2-bromo-2-methylpropane at each time from the start of the reaction. V_{final} is a measure of its initial concentration because each mole of hydrogen bromide is produced from one mole of 2-bromo-2-methylpropane.

2 The half-life of 2-bromo-2-methylpropane is constant, so the reaction is first order.

3 This is an order of reaction with respect to 2-bromo-2-methylpropane because the water is in very large excess.

4 We do not have enough information to decide which mechanism is operating, because we do not know the effect of changing the concentration of the nucleophile (water, or hydroxide ion).

5 Doubling the concentration of hydroxide ions doubles the time for the indicator to change colour. This means that in the same time, the same amount of acid is produced.

6 Increase of hydroxide ion concentration has no effect on the rate of change

of concentration of 2-bromo-2-methylpropane.
7 The order of the reaction with respect to hydroxide ions is zero.
8 The S_N1 mechanism operates in this case, since the reaction is first order overall.

Summary

At the end of this section, students should:
  1 understand what is meant by a rate equation, rate constant, order of a reaction, and half-life;
  2 realize that the stoicheiometric equation is not a reliable guide to the concentrations that are involved in the rate equation;
  3 have practical experience of the measurements of rates of change of concentrations of substances involved in chemical reactions;
  4 be able to work out an order of reaction from experimental data;
  5 know that it may be possible to obtain some idea of the mechanism of a reaction from a study of the kinetics of the reaction;
  6 know what is meant by a rate-determining step in a reaction mechanism.

14.3
THE EFFECT OF TEMPERATURE ON THE RATE OF A REACTION

Objectives

  1 To provide experimental experience of rate measurements at different temperatures.
  2 To show how the kinetic theory of gases provides a theoretical approach to activation energy, using the collision theory of reaction kinetics.
  3 To introduce the Arrhenius equation empirically.

Timing

Experiment 14.3 will require a double period, the discussion of the collision theory another double period, and some time will be needed for graph drawing and calculation.

Suggested treatment

Students should begin by carrying out Experiment 14.3. If they have previously followed the Revised Nuffield Chemistry topic on rates of reaction the basic idea will be familiar to them. This experiment however is a rather more sophisticated version of the practical technique that they may have used earlier.

EXPERIMENT 14.3
The effect of temperature on the rate of the reaction between sodium thiosulphate and hydrochloric acid

Each student or pair of students will need:
Beaker, 400 cm^3
2 boiling-tubes, 150 × 25 mm
Bunsen burner
Piece of paper with dark ink spot
Adhesive tape to fix paper to beaker
Thermometer, 0–110 °C
2 measuring cylinders, 10 cm^3
Watch or clock reading seconds
0.1M sodium thiosulphate solution, 100 cm^3
0.5M hydrochloric acid, 100 cm^3

Procedure

Full details are given in the *Students' book*. The following points should be emphasized when introducing this experiment.

1 Two measuring cylinders should be used. They should be labelled, one for each solution, and the two should not be confused. If they are, sulphur precipitates will start to form in them, making the experimental results invalid.

2 Similarly, the test-tubes should be labelled, A and B, so that the same one can be kept for the same solution in each run of the experiment.

3 Test-tube A should be rinsed thoroughly with water between runs. If this is not done, the acid left over from the previous run will react with the new portion of sodium thiosulphate solution whilst it is being brought up to the required temperature.

Sample results are shown on the graph, figure 14.2. This experiment is adapted from TURNER, J.W., 'The activation energy of the thiosulphate–acid reaction'. *School Science Review* **53**, *185*, p. 751 (1972).

Teachers should note that when completing the table, students enter the time, units seconds, and then work out the rate by taking the reciprocal of the time, units seconds^{-1}. Finally they enter the logarithm of the rate, ln(rate). Strictly speaking one can only take logarithms of dimensionless quantities; one cannot take logarithms of quantities with units. To get around this difficulty, the trick is to divide each rate by 1 second^{-1}, thus preserving the numerical value but removing the units. To be precise, therefore, the final column in the table in the *Students' book* should be headed

$$\ln \frac{\text{rate}/\text{s}^{-1}}{1/\text{s}^{-1}}$$

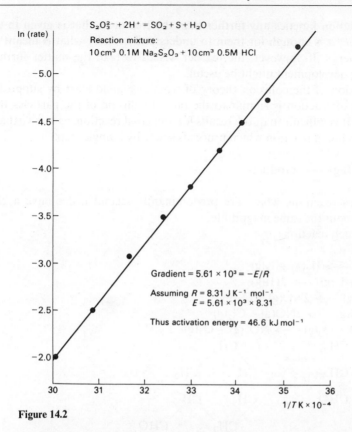

Figure 14.2

Teachers may wish to discuss this with their more mathematically-minded students, but it has not been mentioned in the *Students' book*. The same point arises several times in Topic 15, where logarithms of concentrations figure prominently, and it is solved in the same way; each concentration, with units mol dm^{-3}, being divided by 1 mol dm^{-3}. Again, the *Students' book* does not mention the problem.

The collision theory of reaction kinetics

In the discussion of the collision theory of reaction kinetics in the *Students' book*, teachers should note that the gas constant, R, is used in place of Lk, the product of the Avogadro and Boltzmann constants. This is done so as to avoid confusion between the rate constant and the Boltzmann constant, both of which are normally represented by k, and which would otherwise both appear in the same equation. This is explained in the *Students' book*.

It is not intended that students should take the study of the collision theory of reaction kinetics any further than the treatment that is given in the *Students' book*; it is enough for them to understand broadly what is meant by activation energy. If, however, the teacher wishes to take the matter further, the following development might be useful.

Discussion of the collision theory of reactions could start by supposing that the rate of reaction is similar to the rate of collision of the particles that are reacting. It is difficult to quote results for an actual reaction, as it is virtually impossible to find a reaction which proceeds solely by a single step:

$A(g) + B(g) \longrightarrow$ products

but many gas reactions which are predominantly second order have a rate constant of about the same magnitude.

Some such reactions are:

$2HI(g) \longrightarrow H_2(g) + I_2(g)$
$H_2(g) + I_2(g) \longrightarrow 2HI(g)$
$2NO_2(g) \longrightarrow 2NO(g) + O_2(g)$
$2NOCl(g) \longrightarrow 2NO(g) + Cl_2(g)$
$NO_2(g) + O_3(g) \longrightarrow NO_3(g) + O_2(g)$

$$\begin{array}{c} CH_2 \\ \parallel \\ CH \\ | \\ CH \\ \parallel \\ CH_2 \end{array} + \begin{array}{c} CH_2 \\ \parallel \\ CH \\ | \\ CHO \end{array} \longrightarrow \begin{array}{c} CH_2 \\ / \quad \backslash \\ CH \quad CH_2 \\ \parallel \quad \quad | \\ CH \quad CH \\ \backslash \quad / \quad \backslash \\ CH_2 \quad CHO \end{array}$$

We will therefore consider a hypothetical gas reaction of the type

$2A(g) \longrightarrow$ products,

where A has a relative molecular mass of 125, taking place at 773 K with an initial value for $[A(g)]$ of 10^{-2} mol dm^{-3}, for which $k = 10^{-2}$ dm^3 mol^{-1} s^{-1}.

Two pieces of information are needed:
the total number of collisions per second in a given volume;
the rate of reaction in terms of the number of effective collisions per second in the same volume.
The common volume will be taken as one cubic metre.

The total number of collisions can be found by considering a cylinder of twice the average diameter of the molecules of A, and v metres long. A molecule travelling at v m s^{-1} will collide with all the particles whose centres are within this cylinder during one second (see figure 14.3).

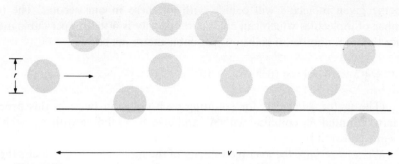

Figure 14.3

The average radius of the molecules and their average velocity can be found. The total number of molecules within the cylinder, which equals the number of molecules with which the chosen particle will collide per second, can be calculated from the original concentration of the gas. The result can be multiplied to account for all the particles, a process in which each collision must be considered twice, so the answer must be divided by two to allow for this, and a value for the total number of collisions found.

For molecules of radius 0.4 nm, with relative molecular mass 125 and average velocity 3.6×10^2 m s^{-1} at 773 K, *the total number of collisions per cubic metre per second* $= 3.2 \times 10^{33}$.

Semi-rigorous derivation, if required

A collision between two identical molecules will occur if their centres approach within a distance equal to their average diameter. This distance can be found from gas viscosity measurements. A molecule of diameter 0.4 nm, travelling at a speed of v m s^{-1} will cover v metres in one second and during this second will collide with every molecule whose centre is within a cylinder of length v cm and radius 0.4 nm.

The volume of this cylinder ($\pi r^2 l$) is $3.14(0.4 \times 10^{-9})^2 \, v$ m^3 = $5 \times 10^{-19} \, v$ m^3.

When $[A(g)] = 10^{-2}$ mol dm^{-3} the total number of molecules available *per cubic metre* for collision with the given molecule is

$10^{-2} \, L \times 10^3$ (L, Avogadro constant $= 6 \times 10^{23}$ mol^{-1}; 1 m$^3 = 10^3$ dm^3)
$= 10^{-2} \times 6 \times 10^{23} \times 10^3 = 6 \times 10^{24}$

Thus the number of molecules within the cylinder

$= 6 \times 10^{24} \times 5 \times 10^{-19} v = 3 \times 10^6 v$

and the given molecule will collide with all these in one second. The total number of molecules which can collide in this way is 6×10^{24} per cubic metre, hence the total number of collisions per cubic metre per second

$$= \tfrac{1}{2}(3 \times 10^6 \, v)(6 \times 10^{24}) = 9 \times 10^{30} \, v$$

(The factor $\tfrac{1}{2}$ corrects for counting each collision twice in this process; we have counted A_1 colliding with A_2 and also A_2 colliding with A_1, which is the same collision.)

We now require the average velocity of the molecules, from kinetic theory of gases; this is given by

$$v = \left(\frac{8 \times 10^3 \, RT}{\pi M} \right)^{\tfrac{1}{2}}$$

where M = relative molecular mass
and $R = 8.3 \, \text{J mol}^{-1} \text{K}^{-1}$ (the gas constant).
At 773 K with $M = 125$

$$v = \left(\frac{8 \times 10^3 \times 8.3 \times 773}{3.14 \times 125} \right)^{\tfrac{1}{2}} = 3.6 \times 10^2 \, \text{m s}^{-1}$$

The total number of collisions per cubic metre per second

$$= 9 \times 10^{30} \times 3.6 \times 10^2 = 3.2 \times 10^{33}$$

(*Note.* In this treatment some assumptions have been made about the velocity of the given molecule, for example that it does not deviate from a straight line path throughout the v metres of its travel. A more rigorous treatment gives a value of 3.6×10^{33} collisions m^{-3} s^{-1}.)

The rate of the reaction in terms of the number of effective collisions (those which lead to reaction) can be calculated quite easily. For $[A(g)] = 10^{-2}$ mol dm^{-3} and $k = 10^{-2}$ dm^3 mol^{-1} s^{-1} it is *number of effective collisions per cubic metre per second* $= 3 \times 10^{20}$.

Rigorous derivation, if required

Rate of reaction $= k[A(g)]^2$
$$= 10^{-2} \times (10^{-2})^2 = 10^{-6} \, \text{mol dm}^{-3} \text{s}^{-1}$$

number of molecules decomposing per second per cubic decimetre

$$= 10^{-6} L = 10^{-6} \times 6 \times 10^{23}$$

but $1 \text{ m}^3 = 10^3 \text{ dm}^3$

∴ number of molecules decomposing per second *per cubic metre*

$$= 10^{-6} \times 6 \times 10^{23} \times 10^3$$
$$= 6 \times 10^{20}$$

Two molecules are required to effect one collision.

∴ number of effective collisions per cubic metre per second

$$= \tfrac{1}{2} \times 6 \times 10^{20} = 3 \times 10^{20}$$

Thus only a minute percentage $\left(\dfrac{3 \times 10^{20} \times 100}{3.2 \times 10^{33}} \approx 10^{-11} \text{ per cent} \right)$ of the collisions result in reaction. Let us consider the reaction more closely. When collision occurs, what happens next? Whether the reaction is a decomposition or any other type, normally the first occurrence is the breaking of bonds. Other bonds may be made later and the overall energy may be exothermic or endothermic, but before other bonds can be made, some must be broken. Obviously this initial breaking of bonds requires energy and therefore only those collisions, after which the particles have sufficient energy for these bonds to be broken, will result in reaction. The fact that certain initial minimum energy is required is demonstrated by gunpowder, coal, a match and many other examples of exothermic reactions which nevertheless require an initial input of energy.

If this is right, it becomes important to find what fraction of the collisions have this certain minimum energy. First, what value should be used for the minimum energy? Before the bond is completely broken, a new bond will have started to form so the energy required would not be quite as large as the average of bond energy terms. These vary from about 150 kJ mol^{-1} for the weak I—I bond to about 550 kJ mol^{-1} for strong bonds such as H—F, with other single bond values intermediate (O—H, 464 kJ mol^{-1}; C—C, 347 kJ mol^{-1}). With the average single bond value at around 350 kJ mol^{-1}, a reasonable guess might be 200 kJ mol^{-1}.

How can the fraction of the collisions that have an energy greater than 200 kJ mol^{-1} be calculated? The energy of the particles is mainly a function of their velocity and the distribution of velocities can be experimentally determined by an apparatus such as that used by Zartman in 1931. Although the velocities of particles are continually changing due to collision, the fraction of particles with a certain velocity remains constant. Zartman's technique can best be explained by means of a film loop: Longman's chemistry film loop no. 8

'Experiments to show Maxwell–Boltzmann distribution of velocities and energies in molecules of gas' is suitable. This film loop is no longer obtainable from the publishers, but will be useful if still available in the school or local teachers' centre. The graph in figure 14.4 shows how the velocities are distributed.

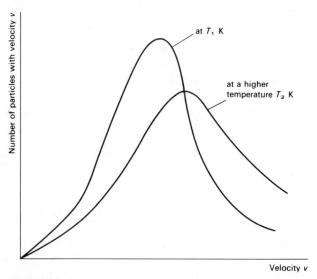

Figure 14.4

This distribution can also be calculated theoretically and the fraction of collisions with an energy greater than a value E J mol^{-1} is given by:

$$\ln\left(\frac{\text{number of collisions with energy } E}{\text{total number collisions}}\right) = -\frac{E}{RT}$$

or

$$\left(\frac{\text{number of collisions with energy } E}{\text{total number collisions}}\right) = e^{-E/RT}$$

The fraction of the molecules with energy greater than 200 kJ mol^{-1} at 773 K is given by

$$\ln(\text{fraction}) = -\frac{E}{RT} = -\frac{200 \times 10^3}{8.3 \times 773}$$
$$= -31.2$$

from which the fraction of collisions with total energy greater than 200 kJ mol^{-1} = 2.9×10^{-14}.

This corresponds to one collision in 4×10^{13}.

Thus the rate at which collisions occur having an energy greater than $200\,\text{kJ mol}^{-1}$

 = total number of collisions per cubic metre per second × fraction with energy $> 200\,\text{kJ mol}^{-1}$
$= 3.2 \times 10^{33} \times 2.9 \times 10^{-14}$
$= 9 \times 10^{19}\,\text{m}^{-3}\,\text{s}^{-1}$

This is not far off the number of effective collisions calculated above, $3 \times 10^{20}\,\text{m}^{-3}\,\text{s}^{-1}$. To make the two results identical a value of $192\,\text{kJ mol}^{-1}$ for E is needed, so the original guess was not too bad.

This minimum energy, E is called the *energy of activation*. The reaction can be imagined as proceeding in something of the manner shown in figure 14.5.

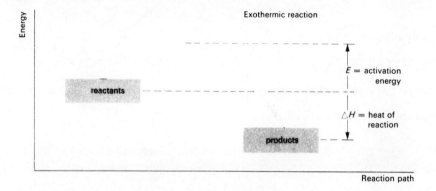

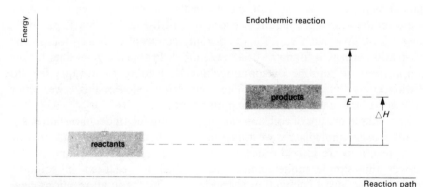

Figure 14.5

Suppose the rate of the reaction depends on temperature according to the relation

rate of reaction $\propto e^{-E/RT}$

The rate constant, k, is a measure of the rate of the reaction, independent of concentration, so

$k \propto e^{-E/RT}$

∴ $k = \text{constant} \times e^{-E/RT}$

∴ $\ln k = \ln \text{constant} - \dfrac{E}{RT}$

since the logarithm of a constant is also constant we have

$\ln k = c - \dfrac{E}{R} \times \dfrac{1}{T}$

Comparing this with

$y = c + mx$

means that a graph of $\ln k$ against $\dfrac{1}{T}$ should, if the theory is correct, be a straight line, and its gradient will be

$-\dfrac{E}{R}$

where E = activation energy in J mol^{-1} and $R = 8.3$ J K^{-1} mol^{-1}. Students can find out if this is true and then obtain a value for the activation energy by a graphical method. To do this, values of the rate constant for given reactions at different temperatures are required. Problem 7 at the end of this Topic in the *Students' book* gives such data for the decomposition of hydrogen iodide and the activation energy is found to be 183 kJ mol^{-1}. Problem 8 gives data for the decomposition of benzene diazonium chloride, leading to a value for the activation energy of 116 kJ mol^{-1}. The connection between this lower value and the rate of the reaction should be pointed out.

This simple picture of collision can only be a useful model in certain cases. Difficulties arise immediately we think of a first order reaction and, indeed, very few reactions are second order one-step processes as in our simple model. Moreover this simple model cannot be applied to reactions in solution. Nevertheless the idea of activation energy is a useful one in all reactions since it can be obtained experimentally, and need not be derived by application of a collision theory.

Summary

At the end of this section, students should:

1 have some practical experience of measuring the effect of temperature on the rate of change of concentration of substances involved in chemical reactions;

2 know about the collision theory and its connection with the kinetic theory of matter;

3 be aware of the Arrhenius equation, and be able to use it to obtain activation energies of reactions.

14.4
CATALYSIS

Objectives

1 To give students the opportunity to collect together their knowledge of catalysis.

2 To use the Arrhenius equation to show that a catalyst provides a reaction pathway of lower activation energy.

3 To provide an experiment illustrating autocatalysis.

Timing

One double period should be sufficient for the theory and another for the practical work. The graph drawing and deductions will probably take another single period.

Suggested treatment

The teaching of catalysis is difficult to place in any chemistry course because if all the ideas are to be dealt with at one time, it must be left as late as possible. If the suggested order of topics has been followed, students will not so far have studied electrode potentials or the transition elements. It is not therefore possible for them to complete their study of catalysis at this point, nor should they be left with the impression that this is the only place in the course where the subject is discussed.

The principal advance that can be made at this stage is the introduction of the idea that catalysts provide reaction pathways of lower activation energy than that of the uncatalysed reaction. One example of this is given in detail in the *Students' book*; other examples of activation energies for catalysed and uncatalysed reactions are given in the *Book of data*.

EXPERIMENT 14.4
A kinetic study of the reaction between manganate(VII) ions and ethanedioic acid

This reaction involves 'autocatalysis' though there is no mention of this word in the *Students' book*. The experiment can therefore be carried out in a spirit of investigation, both to find out what happens and to try to explain the results. This is therefore a good experiment to use as an assessed practical in which *interpretation* is assessed. Again, two methods are offered.

Method 1

Each student or pair of students will need:
Flat-bottom flask or other container, 500 cm^3, to use as a reaction vessel
Conical flask, 100 cm^3, for titration
2 measuring cylinders, 100 cm^3
Measuring cylinder, 10 cm^3
Burette, 50 cm^3, and burette stand
Pipette, 10 cm^3, and safety filler
0.2M ethanedioic acid (poisonous), 100 cm^3
0.2M manganese(II) sulphate, 15 cm^3 (for about half the students)
2M sulphuric acid, 5 cm^3
0.02M potassium manganate(VII), 50 cm^3
0.01M sodium thiosulphate, 200 cm^3
1% starch solution, 15 cm^3
0.1M potassium iodide, 100 cm^3

Procedure

Full details are given in the *Students' book*. It is suggested that half of the class should do Experiment 1 and the other half Experiment 2. Results should then be exchanged so that graphs from both experiments can be drawn by everyone.

Results

As a result of the experiments of Method 1, the students should obtain graphs of the form shown in figure 14.6.

Answers to the questions given after Method 1 in the Students' book.
  1 A measure of the concentration of the manganate(VII) ion is given by the titre of sodium thiosulphate, since the manganate(VII) ions react with iodide ions releasing iodine quantitatively, and this in turn reacts quantitatively with thiosulphate ions.
  2 The mixture for Experiment 2 contains manganese(II) ions. Initially there are no such ions in the mixture for Experiment 1, but manganese(II) ions are amongst the products of the reaction which takes place. Initially the rate of

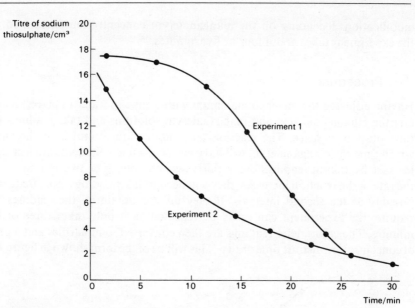

Figure 14.6

change of concentration of manganate(VII) ions is slow, but it accelerates when the manganese(II) ion concentration increases. Manganese(II) ions must be a catalyst for this reaction.

Method 2

In this method the students follow the extent of the reaction with time by measuring the concentration of manganate(VII) ions at suitable intervals using a colorimeter (see Appendix 2).

Each student or pair of students will need:
1 stopclock, or sight of a large clock with seconds hand
1 test-tube
1 colorimeter with tubes (or cuvettes) to fit

Access to communal burettes containing:
Solution which is 0.1M with respect to ethanedioic acid and 1.2M with respect to sulphuric acid
0.02M potassium manganate(VII)
0.02M manganese(II) solution
Carbon dioxide generator

The concentration of manganate(VII) may need adjustment for the particular colorimeter used. The ethanedioic acid concentration may require modification depending on the manganate(VII) concentration chosen, so that the experiment takes about four or five minutes.

Procedure

Having adjusted the meter to maximum with a tube of water in place, students mix the ethanedioic acid and manganate(VII) solutions and take readings with this solution in place. The reaction takes about four minutes, at the end of which time the manganate(VII) will have been consumed. Students are instructed to start by taking readings every thirty seconds but after two minutes, when the rate of the reaction increases, they will need to take readings more frequently, possibly at ten second intervals. If they do not obtain all the readings they require, the experiment can easily be repeated as it only takes three or four minutes. The colorimeter readings are then converted to molarities and a graph of concentration against time drawn. This will be of the form shown in figure 14.7.

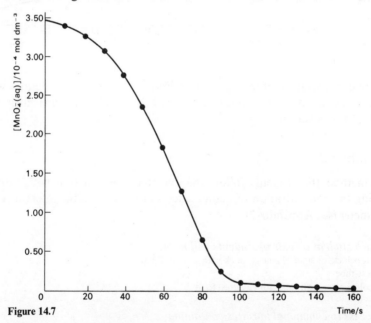

Figure 14.7

Teachers may think it useful for students to take gradients from this graph, and to plot the rates obtained in this way against time. As the concentrations of manganate(VII), ethanedioic acid, and hydrogen ions upon which one might expect the rate to depend are decreasing with time, the rate of the reaction must depend upon the concentration of something which is *increasing* in concentration with time, that is, something which is produced in the reaction. Students will probably need help before they suggest that catalysis by something produced in the reaction is occurring.

Looking at the stoicheiometric equation,

$$2MnO_4^-(aq) + 16H^+(aq) + 5C_2O_4^{2-}(aq) \longrightarrow$$
$$2Mn^{2+}(aq) + 8H_2O(l) + 10CO_2(g)$$

there are two obvious candidates, $Mn^{2+}(aq)$ and $HCO_3^-(aq)$ from dissolved carbon dioxide. Other possibilities are intermediates that may be produced in the reaction such as manganese in other oxidation states. Students can investigate these possibilities by repeating the experiment, adding 1, 2, 4, or 6 drops (approximately 0.02, 0.04, 0.08, or 0.12 cm^3) of 0.2M manganese(II) solution or by saturating the ethanedioic acid solution with carbon dioxide before the experiment. They will find that the addition of manganese(II) moves the $[MnO_4^-]$ against time graph to the left, that is, moving towards a more 'normal' graph. If enough manganese(II) is added the increase in rate will not occur at all. This indicates that manganese(II) is catalysing the reaction.

Whichever method is used, students should finally be introduced to the term 'autocatalysis', which is used to describe this type of reaction.

Students can then be asked how they think the rate of the reaction between iodine and propanone varies with time, if one starts with propanone and iodide ions only. As hydrogen ions are produced, it is also autocatalytic and it would be expected that the form of the extent of reaction/time graph would be similar to that formed for the manganate(VII)-ethanedioic acid reaction.

Supporting material

ICI Videotape 'Catalysis' (colour, sound, 20 minutes) available from Argus Film and Video Library, 15 Beaconsfield Road, London NW10 2LE.

Summary

Students should now have some organized knowledge of catalysis, and understand that catalysts provide a reaction pathway of lower activation energy than is the case in their absence. They should also have some practical experience, and understanding, of autocatalysis.

APPENDICES

This Topic of the *Students' book* ends with two appendices, which are included for the student who may be curious about the mathematical basis of the rate equations, and their relationship with the orders of reactions.

In Appendix 1, rate equations for first and second order reactions are integrated, to give the variation of concentration with time for each case.

Appendix 2 shows how orders of reaction may be obtained from half-life times.

Students should not be expected to learn the contents of these appendices.

ANSWERS TO PROBLEMS IN THE *STUDENTS' BOOK*

(A suggested mark allocation is given in brackets after each answer.)

1a Dilatometry – the very small change of volume of the reaction mixture with time. (2)
 b Measure the volume of gas produced with time, keeping the temperature constant. (2)
 c Withdraw samples of the reaction mixture, quench by cooling, and titrate with standard alkali. (3)
 d Withdraw samples of reaction mixture, quench by cooling, and titrate with standard alkali; a 'final' titration is necessary. (4)
 e Measure the rotation of polarized light as it changes with time. (2)
 f Withdraw samples of reduction mixture, quench by neutralizing the acid with sodium hydrogencarbonate, and titrate with sodium thiosulphate solution; a 'final' titration is necessary. (5)
 g Withdraw samples, quench by neutralizing the acid with sodium hydrogencarbonate, add excess potassium iodide, and titrate the iodine with sodium thiosulphate; a 'final' titration is necessary. (6)

Total 24 marks

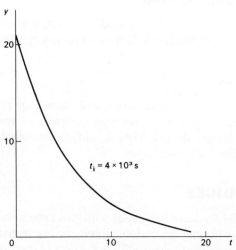

Figure 14.8

2ai The graph should be as shown in figure 14.8. (5)
 ii Since $t_{\frac{1}{2}}$ is constant with a value of 4×10^3 s, the reaction is first order. (3)
 iii This is the order with respect to hydrogen peroxide. (1)
2bi The graph should be as shown in figure 14.9. (5)

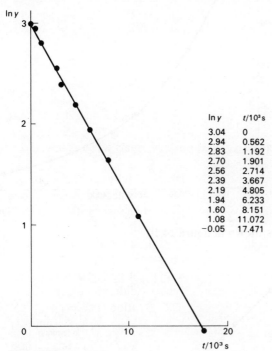

ln y	$t/10^3$ s
3.04	0
2.94	0.562
2.83	1.192
2.70	1.901
2.56	2.714
2.39	3.667
2.19	4.805
1.94	6.233
1.60	8.151
1.08	11.072
−0.05	17.471

Figure 14.9

 ii Slope $= -\dfrac{1.74}{10^4}\,\text{s}^{-1}$
 so $k = 1.74 \times 10^{-4}\,\text{s}^{-1}$ (3)
 and since the graph is a straight line the reaction is first order. (2)

Total 19 marks

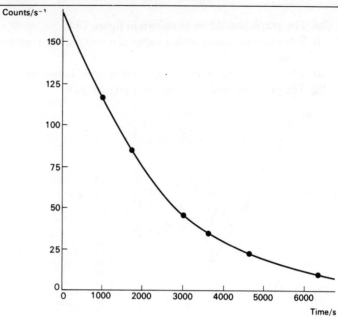

Figure 14.10

3a i The graph should be as shown in figure 14.10. (5)
  ii 1400 s (2)

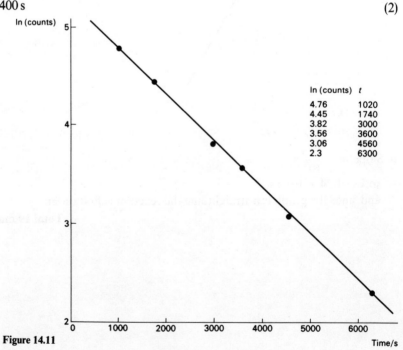

ln (counts)	t
4.76	1020
4.45	1740
3.82	3000
3.56	3600
3.06	4560
2.3	6300

Figure 14.11

b i The graph should be as shown in figure 14.11. (5)
ii $k = 4.6 \times 10^{-4}\,\text{s}^{-1}$ (2)

Total 14 marks

4a It is probably not necessary to quench the reaction in this case because the reaction is already very slow (e.g. 30 minutes between the first and the second readings). (2)
b The titres are proportional to the concentration of hydroxide ions in the reaction mixture. (1)
c The graph should be as shown in figure 14.12. (5)

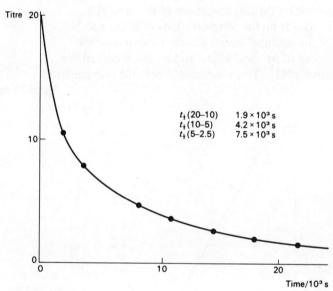

Figure 14.12

Since $t_{\frac{1}{2}}$ values increase sharply with time, the reaction is probably second order. A graph of $\dfrac{1}{\text{titre}}$ which proves to be a straight line would confirm this. (4)
d The S_N2 mechanism is consistent with these results. (1)

Total 13 marks

5a i Rate of formation of iodine = $k[H_2O_2]^1[I^-]^1[H^+]^0$ (3)
ii 1.75×10^{-2} (2)
iii $\text{dm}^3\,\text{mol}^{-1}\,\text{s}^{-1}$ (2)

b Both H_2O_2 and I^- are reactants in the rate-determining step and in such numbers of particles that the reaction should be first order with respect to each of these. H^+ is involved in the mechanism after the rate-determining step so the reaction should be zero order with respect to H^+. This mechanism is therefore consistent with the results. (5)

Total 12 marks

6 a i Rate of formation of bromine $= k[BrO_3^-]^1[Br^-]^1[H^+]^2$ (3)
ii The units of rate constant here would be $dm^9 \, mol^{-3} \, s^{-1}$ (2)
b Considering the rate-determining step,
[HBr] depends on the concentrations of Br^- and H^+;
[HBrO_3] depends on the concentrations of BrO_3^- and H^+.
The rate of the reaction should therefore depend on the concentrations of Br^- and BrO_3^- and on the square of the concentration of H^+. This is consistent with the rate equation. (5)

Total 10 marks

7

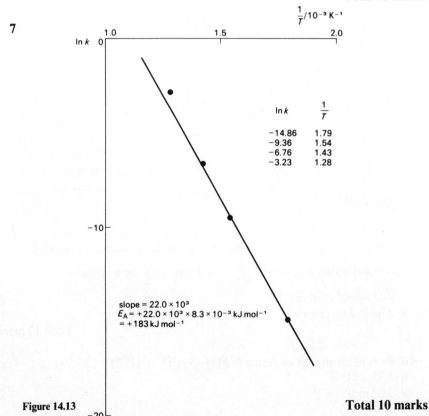

ln k	$\frac{1}{T}$
−14.86	1.79
−9.36	1.54
−6.76	1.43
−3.23	1.28

slope $= 22.0 \times 10^3$
$E_A = +22.0 \times 10^3 \times 8.3 \times 10^{-3} \, kJ \, mol^{-1}$
$= +183 \, kJ \, mol^{-1}$

Figure 14.13

Total 10 marks

8

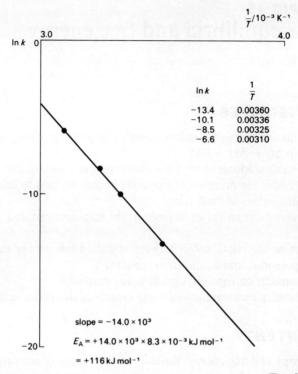

Figure 14.14

slope = -14.0×10^3

$E_A = +14.0 \times 10^3 \times 8.3 \times 10^{-3}$ kJ mol^{-1}

= $+116$ kJ mol^{-1}

Total 10 marks

TOPIC 15
Redox equilibria and free energy

OBJECTIVES

1 To show the meaning and usefulness of free energy changes, ΔG, and of the relationship $\Delta G = \Delta H - T\Delta S$.
2 To enable students to use tabulated values of standard free energy changes.
3 To explain the meaning of electrode potentials, and to enable students to use tabulated values of these data.
4 To provide practical experience in the measurement and use of electrode potentials.
5 To show the relationship between standard free energy change, standard electrode potential, and equilibrium constant.
6 To consider entropy changes in redox reactions.
7 To develop an awareness of some aspects of energetics in life processes.

CONTENT

15.1 Entropy and free energy. Revision of knowledge of entropy; second law of thermodynamics; ΔS_{total} is zero at equilibrium; free energy changes.
15.2 Redox equilibria: metal/metal ion systems. Oxidation and reduction by electron transfer; measurement of p.d. of cells; cell diagrams; the hydrogen electrode; electrode potentials.
15.3 The effect of concentration changes on electrode potentials. Experimental investigation of the effect in the case of silver; standard electrode potentials; the Nernst equation; use of e.m.f. measurements to estimate small concentrations of ions. Background reading: 1 'The pH meter'.
15.4 Redox equilibria extended to other systems. Redox reactions other than metal/metal ion systems; concentration effects in ion/ion systems; use of standard electrode potentials, including prediction of likely reactions, and finding equilibrium constants.
15.5 Entropy considerations. Entropy in cells; free energy and the e.m.f. of cells; finding cell e.m.f. under non-standard conditions.
15.6 Predicting whether reactions will take place; $\Delta G^\ominus$, $E^\ominus_{cell}$, and K_c. The use of standard free energy changes, standard e.m.f. of cells, and equilibrium constants in predicting whether reactions are feasible; relationship between these quantities.
15.7 Standard free energies of formation, and their use. Use of tabulated values

of $\Delta G_f^\ominus$; calculation of equilibrium constants. Background reading: 2 Energetics in life processes.

TIMING
Four weeks will be needed for this Topic.

INTRODUCTION
This Topic sees the fulfilment of all the foundation work that has been done on entropy and free energy. It begins by gathering together the conclusions that have been reached in those sections of previous Topics that have been concerned with this subject. Redox equilibria are then studied so that electrode potentials can be drawn into the discussion. Next, the value and use of standard free energy changes are explained. The relationship between standard free energy changes, standard electrode potentials, and equilibrium constants is then established, and their value in determining the direction of chemical changes and the feasibility of chemical reactions is discussed.

This is an important Topic which should not be hurried; there is much to be gained from a mastery of its contents. Students will then go on to the study of the transition elements, in Topic 16, putting their new-found experience to the test with confidence and understanding.

15.1
ENTROPY AND FREE ENERGY

Objectives

1 To review some of the conclusions arrived at earlier in the course, in Topics 3, 4, 6, 10, and 12.

2 To emphasize, once again, that a system will not undergo a spontaneous change unless the overall entropy change, ΔS_{total}, is positive. Even when this requirement is met there may be other inhibiting factors, such as a high activation energy (see sections 12.6 and 14.3 of this book).

3 To consider a change, that of water vapour in the air of a warm room condensing on a cold window pane, for which ΔS_{system} is negative but $\Delta S_{surroundings}$ is positive. Since $\Delta S_{surroundings}$ is greater than ΔS_{system} the change takes place spontaneously.

4 To show the usefulness of the expression

$$\Delta G = \Delta H - T\Delta S_{system}$$

as a guide to whether a change in a system is feasible.

Timing

Not more than two periods will be required if the students have read through the section beforehand.

Suggested treatment

Again, as in section 12.6, teachers will probably find it most convenient to lead students through the sequence in the *Students' book*, making sure that they appreciate the importance of taking the *surroundings* as well as the *system* into account when a given change is under discussion. It is important also to stress that for a change to be feasible ΔS_{total} must be positive, and that when the system reaches equilibrium, ΔS_{total} is zero.

The meaning of the superscript $\ominus$ should also be stressed, so that when it is used, in the *Book of data* for example, all numerical values of $\Delta H^{\ominus}$, $\Delta G^{\ominus}$, and $S^{\ominus}$ (usually referred to in speech as ΔH standard, ΔG standard, and S standard) apply only to a temperature of 298 K and a pressure of 760 mmHg, known as *standard conditions*.

The students' understanding of the section could be tested by setting them one or two simple problems, such as the following.

Using information given in the *Book of data*, calculate the values of $\Delta S^{\ominus}_{total}$ and $\Delta G^{\ominus}$ for the following changes:

a ethene(gas) + hydrogen(gas) $\longrightarrow$ ethane(gas)
b copper(solid) + hydrogen chloride(gas) $\longrightarrow$
    copper(I) chloride + hydrogen(gas)
c silicon(solid) + hydrogen(gas) $\longrightarrow$ silane(SiH$_4$)(gas)

Which of these changes would you expect to be feasible under standard conditions?

The answers to these problems are:
a $\Delta S^{\ominus}_{total} = +339 \, \text{J K}^{-1} \text{mol}^{-1}$, $\Delta G^{\ominus} = -101 \, \text{kJ mol}^{-1}$; reaction feasible.
b $\Delta S^{\ominus}_{total} = +76 \, \text{J K}^{-1} \text{mol}^{-1}$, $\Delta G^{\ominus} = -25 \, \text{kJ mol}^{-1}$; reaction feasible.
c $\Delta S^{\ominus}_{total} = -191 \, \text{J K}^{-1} \text{mol}^{-1}$, $\Delta G^{\ominus} = +57 \, \text{kJ mol}^{-1}$; reaction not feasible.

Supporting homework

Reading the section before it is discussed in class.

Summary

At the end of this section, students should:

1 have consolidated their understanding of some of the conclusions reached in sections 3.7, 4.5, 6.6, 10.1, 10.2, and 12.6;
2 be able to calculate $\Delta S^{\ominus}_{system}$ and $\Delta S^{\ominus}_{surroundings}$ for a proposed change and hence to decide, from $\Delta S^{\ominus}_{total}$, whether it is likely to take place under standard conditions;
3 understand that when equilibrium is reached ΔS_{total} is zero;
4 appreciate the value of using ΔG, rather than ΔS_{total}, as a guide to feasibility and attainment of equilibrium, since it includes ΔS_{total}, and is often easy to measure, or to calculate from tables of data.

15.2
REDOX EQUILIBRIA: METAL/METAL ION SYSTEMS

Objectives

1 To consider the importance to metal activity of relative tendency to form ions in solution.
2 To deal with metal/metal ion systems as redox equilibria.
3 To introduce the hydrogen electrode.
4 To establish the technique for measuring the e.m.f. of cells and to discuss the information obtainable from such measurements.

Timing

About eight periods. Students who have followed Revised Nuffield Chemistry before starting this course may be familiar with some of the ideas dealt with in this section, which appear in Stage II, Topic A23. If so, less time may be needed.

Suggested treatment

Redox systems have been studied earlier in the course, especially in Topic 5, where a change of oxidation number was used as a criterion for redox processes. Here we deal with the electron transfers which take place in redox reactions. It is suggested that a simple class experiment on two displacement reactions

$$Zn(s) + Cu^{2+}(aq) \longrightarrow Zn^{2+}(aq) + Cu(s)$$

and $\quad Cu(s) + 2Ag^{+}(aq) \longrightarrow Cu^{2+}(aq) + 2Ag(s)$

be used to provoke discussion.

EXPERIMENT 15.2a
Some simple redox reactions

Each student or pair of students will need:
4 test-tubes, 100 × 16 mm
Copper foil, 12 × 1 cm
Copper powder, about 0.5 g
Zinc foil, 12 × 1 cm
Zinc powder, about 0.5 g
0.5M copper(II) sulphate solution
0.1M silver nitrate solution
Thermometer, −10 to +110 °C, × 1°

Procedure

Details are given in the *Students' book*. Results to be expected are as follows.

1 Zinc foil dipped in copper(II) sulphate solution will acquire a coating of metallic copper.

$$Zn(s) + Cu^{2+}(aq) \longrightarrow Zn^{2+}(aq) + Cu(s)$$

2 Zinc powder reacts more quickly with the copper(II) solution, as a greater surface area of the metal is in contact with the solution.

3 and 4 Similar results are obtained with copper and silver nitrate solution:

$$Cu(s) + 2Ag^{+}(aq) \longrightarrow 2Ag(s) + Cu^{2+}(aq)$$

Heat is evolved during the reactions, indicating that the energy content of the products is less than that of the reactants in each case.
In (1) and (2)

and
Zn(0) is oxidized to Zn(II)
Cu(II) is reduced to Cu(0)

In (3) and (4)

and
Cu(0) is oxidized to Cu(II)
Ag(I) is reduced to Ag(0)

Oxidation and reduction by electron transfer

Discussion of the questions posed at the end of the experiment should lead to the establishment of the following points.

a Redox processes can involve transfer of electrons. Loss of electrons is oxidation; gain of electrons is reduction.

b The terms oxidant (or oxidizing agent) and reductant (reducing agent) are relative only. In the experiment copper is seen to be capable of acting in both senses.

c The equation for a redox process can be split into 'half-equations' so that the electron transfer is made plain, e.g.

$$Zn(s) \longrightarrow Zn^{2+}(aq) + 2e^- \quad \textit{oxidation}$$
$$2e^- + Cu^{2+}(aq) \longrightarrow Cu(s) \quad \textit{reduction}$$

As these processes are seen to be reversible, they can be treated as equilibria

$$Zn(s) \rightleftharpoons Zn^{2+}(aq) + 2e^-$$
and $$2e^- + Cu^{2+}(aq) \rightleftharpoons Cu(s)$$

Application of Le Châtelier's principle to such equilibria tells us that the relative tendency of the two metals to form ions in solution determines the outcome of a given reaction.

d During a metal/metal ion reaction energy is transferred from system to environment.

e It is very important at this stage to make it clear that the energy transfer during the change:

$$Zn(s) \longrightarrow Zn^{2+}(aq) + 2e^-$$

shown above is *not* the same as the ionization energy of zinc:

$$Zn(g) \longrightarrow Zn^{2+}(g) + 2e^-$$

The formation of ions in solution is quite different from the conversion of gaseous atoms of a metal into gaseous ions.

Measuring the tendency of a metal to form ions in solution

The possibility of using metal/metal ion reactions as sources of electrical energy in a voltaic cell arises from the half-equations considered in interpreting Experiment 15.2a. A voltaic cell provides a means of comparing the relative tendencies of metal/metal ion couples to liberate electrons by forming ions in solution. *Absolute* electrode potentials cannot be measured, but potential *differences* can be measured easily.

Experiment 15.2b explores the variation of potential difference between the electrode systems in a Daniell cell with the change of resistance in a circuit. This enables the idea of the electromotive force (e.m.f.) of a cell as the maximum p.d. obtainable, to be introduced. The symbol E is used for e.m.f. This experiment should be preceded by a brief description of the Daniell cell, including the function of the porous pot in preventing admixture of the electrode solutions whilst allowing electrical contact between them.

EXPERIMENT 15.2b
The variation of p.d. in a Daniell cell with change of external resistance

Each student or pair of students will need:
Daniell cell
Rheostat (1000 Ω or more)
Ammeter (0–1 A range)
Voltmeter, of high resistance. A transistorized or solid-state instrument is preferable
Switch
6 pieces of connecting wire, fitted with terminals suitable for connecting up the circuit shown in figure 15.1

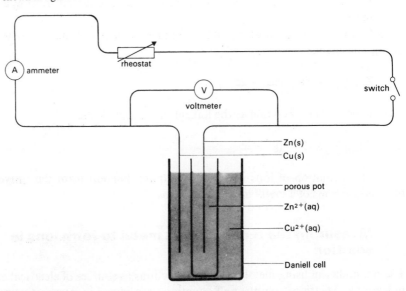

Figure 15.1

Procedure

Full details of the procedure, and a discussion of the results, are given in the *Students' book*.

The observations show that the p.d. increases to a maximum value as the current falls to zero. The maximum p.d. is called the electromotive force of the cell (E volt).

Taking a voltmeter measurement across the cell terminals, switch open, involves a flow of current through the voltmeter, so that an accurate E value cannot be obtained in this way unless the resistance of the voltmeter is so high that the current taken is negligible. A transistor or solid-state voltmeter is the best instrument for this purpose. If one is not available, a potentiometer can be used. Details are given in Appendix 2 in this book.

Cell diagrams

The IUPAC convention is commonly used in Britain and Europe. In this the sign before the E value gives the polarity of the *righthand electrode* in the cell diagram. Of itself, of course, the e.m.f. always represents an energy loss to the system so the $+$ or $-$ sign has no connection with gain or loss of electrical energy. Thus for the Daniell cell, the appropriate method of representation can be either

$\text{Zn(s)}|\text{Zn}^{2+}(\text{aq})\vdots\text{Cu}^{2+}(\text{aq})|\text{Cu(s)}; \quad E = +1.1\text{ V}$

or $\quad \text{Cu(s)}|\text{Cu}^{2+}(\text{aq})\vdots\text{Zn}^{2+}(\text{aq})|\text{Zn(s)}; \quad E = -1.1\text{ V}$

The vertical broken line represents the salt bridge or porous partition.

Contributions made by separate electrode systems to the e.m.f. of a cell

Measurement of the potential of a single electrode system is impossible, because two such systems are needed to make a complete cell of which the e.m.f. can be measured. We can, however, assess the *relative* contributions of single electrode systems to cell e.m.f.s by choosing one system as a standard against which all other systems are measured. The standard system is then arbitrarily assigned zero potential and the potentials of all other systems referred to this value. By international agreement the hydrogen electrode has been chosen as the reference electrode for this purpose.

The hydrogen electrode

The construction of a hydrogen electrode, and its role as a reference electrode, are described in the *Students' book*. The students would no doubt be interested to see a hydrogen electrode and to satisfy themselves that a potential difference can actually be obtained from a cell having hydrogen gas as one of its electrodes. If desired the following experiment may be done as a teacher demonstration.

TEACHER DEMONSTRATION
To measure the e.m.f. of simple cells using the hydrogen electrode as the common reference electrode

The teacher will need:

Hydrogen electrode; platinized Pt wire dipping into 1.0M HCl(aq) in small squat-form beaker (50 or 100 cm^3) and hydrogen cylinder (see Appendix 2)
Cu^{2+}(aq)|Cu(s) electrode
Zn^{2+}(aq)|Zn(s) electrode
Ag^+(aq)|Ag(s) electrode
These electrodes should consist of about 6 × 1 cm strips of metal foil, cleaned with fine emery paper before use, dipping into solutions of 1M copper(II) sulphate, 1M zinc sulphate, and 0.1M silver nitrate respectively in small squat-form beakers (50 or 100 cm^3). The metal strips can be supported in small clamps between pieces of cork, and electrical connection made to them by leads with a crocodile clip attached to one end. A copper foil backing, kept well clear of the solution, is useful in enabling a good connection to be made between crocodile clip and silver electrode.
Potassium nitrate salt bridges. These consist of a single strip of filter paper, 10 × 1 cm, soaked in saturated potassium nitrate solution.
Transistor or solid-state voltmeter (pH meter with appropriate millivolt scale is suitable). If one is not available, a potentiometer can be used (see Appendix 2).
2 pieces of connecting wire, fitted with suitable terminals.

Procedure

Set up apparatus as in figure 15.2. Have the hydrogen electrode on the *lefthand side as the class sees the apparatus*. The end of the glass tube surrounding the platinized platinum wire should be immersed in the 1.0M HCl(aq) as near to the surface as possible, so that the gas pressure is very nearly 1 atm. Use the crocodile clips to connect the electrodes to the valve voltmeter. Note that the copper electrode must be connected to the + terminal of the voltmeter.

Adjust the flow of hydrogen to a rate of about one bubble every two seconds. Read the cell e.m.f. on the valve voltmeter.

(*Note.* If the platinum black has been exposed to the air, the hydrogen electrode will take a little time to reach equilibrium.)

This e.m.f. will be the E value for the cell:

$Pt[H_2(g)] | 2H^+(aq) \vdots Cu^{2+}(aq) | Cu(s)$

The experiment can then be repeated with zinc and silver foils in place of

15.2 Redox equilibria: metal/metal ion systems

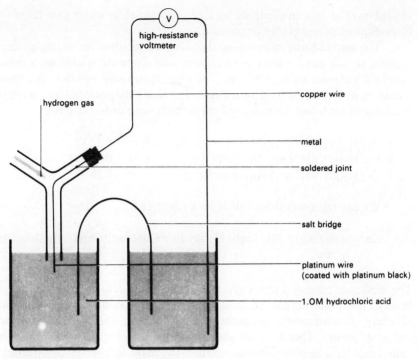

Figure 15.2

the copper foil. Note that in the case of zinc, the hydrogen electrode will have to be connected to the + terminal of the voltmeter; in the case of silver, as of copper, the hydrogen electrode will have to be connected to the − terminal. These e.m.f.s will be the E values for the cells:

$Zn(s)|Zn^{2+}(aq) \vdots 2H^{+}(aq)|[H_2(g)]\, Pt$

$Pt\, [H_2(g)]|2H^{+}(aq) \vdots Ag^{+}(aq)|Ag(s)$

Discussion

Results should be as now indicated:

- **a** $Pt\, [H_2(g)]|2H^{+}(aq) \vdots Cu^{2+}(aq)|Cu(s);$ $E = +0.32\,V$
- **b** $Zn(s)|Zn^{2+}(aq) \vdots 2H^{+}(aq)|[H_2(g)]\, Pt;$ $E = +0.75\,V$
- **c** $Pt\, [H_2(g)]|2H^{+}(aq) \vdots Ag^{+}(aq)|Ag(s);$ $E = +0.76\,V$

The E values obtained give no indication of the absolute potentials that can be attributed to an individual electrode. To do this by a method of this kind we should have to find an electrode with zero potential, in which case there would be no flow of electrons into or out of it.

The simplest way of assessing the relative contributions of single electrode systems to cell e.m.f. values is to choose one electrode system as a reference standard and measure the E values of all other systems against this. The standard chosen is, of course, the hydrogen electrode. Giving the potential of this electrode the value of zero, and rewriting cell (b) in the reverse order we have:

a Pt $[H_2(g)] | 2H^+(aq) \vdots Cu^{2+}(aq) | Cu(s)$; $E = +0.32$ V
b Pt $[H_2(g)] | 2H^+(aq) \vdots Zn^{2+}(aq) | Zn(s)$; $E = -0.75$ V
c Pt $[H_2(g)] | 2H^+(aq) \vdots Ag^+(aq) | Ag(s)$; $E = +0.76$ V

We can represent these values on a linear chart,

$Zn^{2+}(aq) \| Zn(g)$	$2H^+(aq) \| [H_2(g)]$ Pt	$Cu^{2+}(aq) \| Cu(s)$	$Ag^+(aq) \| Ag(s)$
-0.75 V	0 V	$+0.32$ V	$+0.76$ V

The electrode potential series given in the *Book of data* can now be discussed briefly. The point should be made that these $E^\ominus$ values are obtained under carefully specified conditions (hence the superscript) which will be discussed in the next section. Therefore we should not expect exact comparison with the values that have been obtained experimentally in the teacher demonstration. The most important point to stress at this stage is that the E values given are the voltages of *real and complete* cells, of which the *lefthand* electrode is always

$Pt_2[H(g)] | 2H^+(aq) \vdots$ (Hence its position on the bench during the demonstration.)

The order of oxidants and reductants is important in the above chart; it is the order given in the tables of electrode potentials framed according to the IUPAC convention. (It is unfortunate that some countries adopt the opposite convention which entails a reversal of sign for E values, e.g. $+0.75$ for $Zn(s) | Zn^{2+}(aq)$; if old American textbooks are used as reference sources, students should be warned of this discrepancy.)

Experiment 15.2c which now follows is intended to give an opportunity for students to gain experience in constructing voltaic cells and making e.m.f. measurements for themselves.

EXPERIMENT 15.2c
To measure the e.m.f. of some voltaic cells

Each student or pair of students will need:
Copper, zinc, and silver half cells
Potassium nitrate salt bridges
Transistor or solid-state voltmeter
2 pieces of connecting wire

Details of this apparatus are given in the list of apparatus needed for the teacher demonstration, page 100.

Details of alternative cells are given in figure 15.3. Type B in this figure has two advantages:

1 There is a considerable saving of chemicals, especially if the silver nitrate solution is placed in the inner tube.

2 The complete cell is easily portable, hence each student (or pair) can make up one cell, which can be shared with other students.

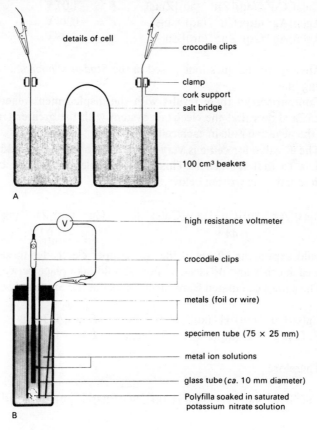

Figure 15.3
Details of alternative cells.

Procedure

Students measure the e.m.f.s of the cells:

a $Cu(s)|Cu^{2+}(aq) \vdots Zn^{2+}(aq)|Zn(s)$
b $Ag(s)|Ag^{+}(aq) \vdots Cu^{2+}(aq)|Cu(s)$
c $Ag(s)|Ag^{+}(aq) \vdots Zn^{2+}(aq)|Zn(s)$

If cell A is used, the salt bridge must be changed for each measurement. Full details are given in the *Students' book*.

Discussion

Results obtained by a sixth form set in this experiment are given below. They are class averages, omitting one result for cell **a** which was obviously incorrect.

a $Cu(s)|Cu^{2+}(aq) \vdots Zn^{2+}(aq)|Zn(s)$; $\quad E = -1.07 \text{ V}$
b $Ag(s)|Ag^{+}(aq) \vdots Cu^{2+}(aq)|Cu(s)$; $\quad E = -0.44 \text{ V}$
c $Ag(s)|Ag^{+}(aq) \vdots Zn^{2+}(aq)|Zn(s)$; $\quad E = -1.53 \text{ V}$

Answers to the questions posed in the *Students' book* should be along the following lines.

1 Comparison of these results with the displacement reactions in Experiment 15.2a shows that the electrode system with the greater tendency to form ions is the negative pole in each cell.

2 The E value for cell **c** is very nearly that obtained by adding the E value for cell **a** to that for cell **b**. This is reasonable because by connecting cells **a** and **b** together as written below

$$Ag(s)|Ag^{+}(aq) \vdots Cu^{2+}(aq)|\overbrace{Cu(s) \quad Cu(s)}|Cu^{2+}(aq) \vdots Zn^{2+}(aq)|Zn(s)$$
$$-0.44 \text{ V} \qquad\qquad\qquad -1.07 \text{ V}$$

we should expect the effects of the two copper electrodes to cancel since they oppose each other and no electron flow should take place between them.

3 The e.m.f.s calculated from the list given in the *Students' book* are:

a $Cu(s)|Cu^{2+}(aq) \vdots H^{+}(aq)|[H_2(g)] \text{ Pt}$; $\quad E = -0.34 \text{ V}$
b $Pt [H_2(g)]|H^{+}(aq) \vdots Zn^{2+}(aq)|Zn(s)$; $\quad E = -0.76 \text{ V}$

Therefore

$$Cu(s)|Cu^{2+}(aq) \vdots H^{+}(aq)|[H_2(g)] \overbrace{Pt(s) \quad Pt(s)}[H_2(g)]|H^{+}(aq) \vdots Zn^{2+}(aq)|Zn(s)$$

in which the two hydrogen electrodes cancel out, will have an e.m.f.

$$-0.34 + (-0.76)\,\text{V} = -1.1\,\text{V}$$

similarly **b** should have an e.m.f. of -0.46 V
and **c** should have an e.m.f. of -1.56 V

Suggestions for homework

Calculating cell e.m.f. for different electrode combinations from information in the table of standard electrode potentials in the *Book of data*. Writing equations for cell reactions. Identifying positive and negative electrodes.

Summary

At the end of this section students should:

1 have some familiarity with the electron transfer aspect of redox reactions, and be able to divide these into half-reactions in simple cases;

2 appreciate that redox reactions can involve equilibria which are governed by the relative tendencies of the half-reactions to liberate electrons;

3 appreciate the fundamental differences between ionization of gaseous atoms, and ion formation in solution from crystal lattices;

4 be able to interpret simple cell diagrams;

5 know the details of the construction of a simple hydrogen electrode;

6 know the use of the hydrogen electrode as a reference point in the conventional tables of electrode potentials;

7 be able to determine experimentally the e.m.f. of simple voltaic cells.

15.3
THE EFFECT OF CONCENTRATION CHANGES ON ELECTRODE POTENTIALS
Objectives

1 To examine the effects of concentration changes on electrode potential values.

2 To define what is meant by a standard electrode potential.

3 To use the relationship between silver ion concentration and E value for the silver electrode to explore solubility differences for sparingly soluble silver salts.

Timing

Four or five periods.

Suggested treatment

Students can begin by carrying out Experiment 15.3a, in which the electrode potential of the silver electrode is measured in solutions of varying silver ion concentration.

EXPERIMENT 15.3a
To investigate the effect of changes in silver ion concentration on the potential of the $Ag^+(aq)|Ag(s)$ electrode

Each student or pair of students will need:
Copper and silver half cells as in Experiment 15.2c. (Solutions required are listed below.)
Potassium nitrate salt bridges as in Experiment 15.2c.
[If cell B is used put $Ag^+(aq)$ in the inner tube: cells of different concentrations can be made up by different students (or pairs) and interchanged for measurements.]
Transistor or solid-state voltmeter

Access to:
1.0M copper(II) sulphate solution, about 40 or 70 cm³ depending on beaker size used
Silver nitrate solution in concentrations of 0.01M, 0.0033M, 0.001M, 0.00033M, and 0.0001M*

Time will be saved if the silver nitrate solutions are prepared beforehand. Otherwise students will need 10 cm³ pipettes and 100 cm³ standard flasks to dilute 0.01M solution. 0.001M and 0.0001M are prepared by progressive dilution; 0.0033M by mixing one volume 0.01M with two volumes water; 0.00033M is prepared similarly from 0.001M.

The cell is set up as indicated in figure 15.3.

Procedure

Using the principle of Le Châtelier the students are asked to predict the effect of concentration change on the equilibrium

$$Ag(s) \rightleftharpoons Ag^+(aq) + e^-$$

---

* The activity coefficient of the most concentrated of these solutions, 10^{-2}M, is 0.90; those of the other concentrations are larger. Thus $\ln [Ag^+(aq)]$ is very close to $\ln a_{Ag^+(aq)}$. For the 10^{-2}M solution $\ln [Ag^+(aq)] = -4.60$ and $\ln a_{Ag^+(aq)} = -4.72$.

15.3 The effect of concentration changes on electrode potentials

Decrease of silver ion concentration increases the tendency for metallic silver to form ions in solution. It has been found that the electrode system in a cell with the greater tendency to form ions is the negative pole in each cell, that is, it has the more negative or less positive electrode potential. Thus, if the tendency for the metal to form ions is increased, the electrode potential would be expected to be more negative or less positive. A similar result would be expected for other metal/metal ion equilibria. This prediction is then tested.

The cell

$$Cu(s)|Cu^{2+}(aq) \vdots Ag^{+}(aq)|Ag(s)$$

is set up as indicated in figure 15.3 and the e.m.f. of the cell measured for various values of $[Ag^{+}(aq)]$. Instructions are given in the *Students' book*.

Although silver electrodes, and solutions of silver nitrate, are expensive, the electrodes last indefinitely, and the solutions used are extremely dilute. Several groups of students can share the same small sample of solution of each concentration; the teacher should emphasize that particular attention must be given to cleanliness of containers and electrodes when using these very dilute solutions. If a silver electrode is being dipped successively into each solution, it should be dipped into distilled water and dried with a tissue between each use, and, for type A cells, fresh salt bridges must be used each time.

The results for this experiment will be needed for the measurement of silver ion concentrations in Experiment 15.3b. They should be tabulated as $[ion]/mol\,dm^{-3}$; $\ln[ion]$; $E/volt$. For a note about taking logarithms of quantities which are not dimensionless, see Topic 14, section 14.3.

Graphs of E against $\ln[ion]$ are drawn. An extended concentration axis permits extrapolation to $\ln[Ag^{+}(aq)] = -42$ for use later. Students may need some help in plotting this graph. The e.m.f.s are all positive and the values of $\ln[ion]$ negative.

In order to find the value of $\ln[Ag^{+}(aq)]$ when $E = 0$ (and the value of E when $\ln[Ag^{+}(aq)] = -42$) the graph will be as shown in figure 15.4.

Answers to the questions posed in the *Students' book* should be along the following lines.

1 The e.m.f. for the reaction is made up of the two electrode potentials

$$Cu(s)|Cu^{2+}(aq) \vdots 2H^{+}(aq)|[H_{2}(g)]\,Pt \quad Pt\,[H_{2}(g)]|2H^{+}(aq) \vdots Ag^{+}(aq)|Ag(s)$$
$$-(+0.34\,V) \qquad\qquad\qquad\qquad +0.80\,V$$
$$e.m.f. = -0.34 + 0.80 = +0.46\,V$$

Decrease of silver ion concentration will increase the tendency of metallic silver to form ions in solution and therefore the potential of the silver electrode would be expected to become more negative or less positive. The value of the cell e.m.f. will therefore become more negative or less positive. This is shown by the graph in figure 15.4.

2 As the graph is a straight line

$$E \propto \ln[\text{ion}]$$

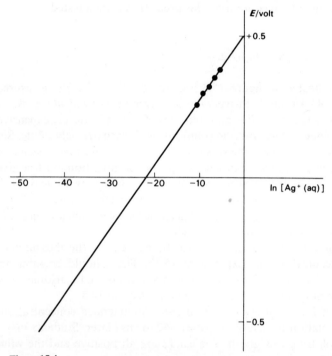

Figure 15.4

3 Using the graph of silver ion concentration against E, the concentration of silver ions for which $E = 0$ can be found. In the cell used $[Cu^{2+}(aq)] = 1$ mol dm^{-3}. This gives the information required to calculate the equilibrium constant for the reaction

$$Cu(s) + 2Ag^+(aq) \rightleftharpoons Cu^{2+}(aq) + 2Ag(s).$$

From a class experiment the value of $\ln[Ag^+(aq)]$ for $E = 0$ was found, by extrapolation, to be -20.25

$$\therefore \quad [Ag^+(aq)] = e^{-20.25} = 1.6 \times 10^{-9}$$

15.3 The effect of concentration changes on electrode potentials

$$\therefore K_c = \frac{[Cu^{2+}(aq)]_{eqm}}{[Ag^+(aq)]^2_{eqm}} = \frac{1}{(1.6 \times 10^{-9})^2} = 4 \times 10^{17} \, dm^3 \, mol^{-1}$$

These experiments show that the value of an electrode potential depends on the concentration of ions in solution. It can also be shown that E varies with temperature. Thus for comparison purposes the temperature and concentration must be standardized. A precise definition of *standard electrode potential* now becomes possible. For our purposes we shall define it as the e.m.f. of a cell in which one electrode is the standard hydrogen electrode and the other electrode consists of a metal in contact with a solution of its ions of concentration $1 \, mol \, dm^{-3}$, the e.m.f. being measured at $25\,°C$. A standard hydrogen electrode consists of hydrogen gas, at a pressure of one atmosphere, bubbling over a platinized platinum surface in a solution which is of concentration $1 \, mol \, dm^{-3}$ with respect to hydrogen ions.

Standard electrode potentials are denoted by the symbol $E^\ominus$ ('E standard' when spoken.) A list is given in the *Book of data*.

A minute or two could be spent in pointing out that the concentration of free ions cannot strictly be equated with the solute concentration as calculated by number of moles of solute in a given volume of solution. Incomplete ionization, inter-ion attraction, and 'crowding effects' in fairly concentrated solution result in the necessity of taking a greater concentration of solute to achieve a given concentration of ions. Thus to obtain a 1.0M solution of free hydrogen ions a concentration of 1.18M hydrogen chloride must be used. No account will be taken of activities or activity coefficients in this Topic but students should be aware that the treatment given here is simplified to some extent.

The information that has been collected so far can be summarized as follows:

1 The standard electrode potential ($E^\ominus$) of an electrode is an intensive property and has a specific value for each electrode system.

2 $E^\ominus$ values are relative, the standard hydrogen electrode being given the arbitrary value of zero. The sign of $E^\ominus$ is the polarity of the metal/metal ion electrode when combined with the standard hydrogen electrode; if it forms the negative pole, the $E^\ominus$ value is given a negative sign, and vice versa.

3 When the ion concentration is other than $1 \, mol \, dm^{-3}$, the E value (not the $E^\ominus$ value) varies with concentration as predicted by the principle of Le Châtelier. E is directly proportional to $\ln[\text{ion}]$. The size of the variation depends on the metal used.

4 The E value is a function of temperature.

The Nernst equation

At temperatures fairly near to 25 °C the variation of E with concentration changes for metal/metal ion systems can be calculated with a fair approximation to accuracy from the relationship

$$E = E^\ominus + \frac{0.026}{z} \ln [\text{ion}]$$

where z = number of charges on the metal ion (change in oxidation number). (The constant has the value of 0.0248 at 15 °C; 0.0256 at 25 °C; 0.0265 at 35 °C.) This fits in with the Le Châtelier prediction of concentration effects since for values of [ion] which are less than 1, $\ln [\text{ion}]$ is negative and the value of E becomes less positive (more negative) with dilution.

The use of graphs obtained in Experiment 15.3a to test this relationship is explained in the *Students' book*. For the silver ion the slope of the E against $\ln [\text{ion}]$ line, which should be 0.026, usually gives quite a good agreement.

The full Nernst equation

$$E = E^\ominus + \frac{RT}{zF} \ln [\text{ion}]$$

for metal/metal ion systems is given in the *Students' book* with a brief explanation of the terms involved. Less able students may make heavy weather of this, so it should not be stressed too much. In case the point is raised, it should be borne in mind that the equation cannot be used to calculate E values at temperatures other than 25 °C, since $E^\ominus$ varies with temperature also. The temperature term relates solely to corrections for varying ion concentrations.

By making use of known variation of ion concentration with E, the concentration of ions in very dilute solutions can be measured electrically. This method has drawbacks with concentrated solutions but can be used for solutions which would be much too dilute to be analysed by ordinary chemical methods. Some examples of this are given in Experiment 15.3b using the copper electrode with 1.0M Cu^{2+}(aq) as a reference electrode in the cell

$Cu(s) | Cu^{2+}(aq) \vdots Ag^{+}(aq) | Ag(s)$

Mixtures of 0.1M silver nitrate solution with various precipitating reagents are used in the silver electrode.

The four parts of the experiment can be shared out amongst members of the class if necessary. The experiment should not occupy more than one double period.

EXPERIMENT 15.3b
Using e.m.f. measurements to estimate small concentrations of silver ions

Each pair of students will need:
Copper and silver half cells as in Experiment 15.2c. (Solutions required are listed below.)
Potassium nitrate salt bridges as in Experiment 15.2c
Measuring cylinder, 25 cm^3
Transistor or solid-state voltmeter

Access to the following solutions in approximately the volumes stated:
1.0M CuSO$_4$(aq), 40 or 70 cm^3, depending on beaker size used
0.1M AgNO$_3$(aq), 30 cm^3
0.1M KCl(aq) or NaCl(aq), 30 cm^3 ⎫
0.1M KBr(aq), 30 cm^3 ⎬ according to allocated reaction
0.1M KI(aq), 30 cm^3 ⎪
0.1M KIO$_3$(aq), 30 cm^3 ⎭
(*Note*. For the alternative cell B much smaller volumes of reagents are adequate.)

Procedure

This is described in the *Students' book*. For most of the mixtures used in the silver electrode system, the reaction taking place reduces the silver ion concentration to such an extent that the sign of E is reversed relative to the reference copper electrode system. Students should be warned of this possibility and told to deal with it by reversing the electrode connections. If this is done, the silver electrode now becomes the *negative pole* of the cell and measured E values must be adjusted accordingly.

From the E values measured for the various electrode combinations, the corresponding values of ln [Ag$^+$(aq)] are obtained from the graph. Values of [Ag$^+$(aq)] are then calculated. Some help may be needed when students change from ln [ion] to [ion]. They should remember that

$$e^{\ln x} = x$$

so, if they have a value for ln [ion] they should enter this into their calculators, press the e^x button, and obtain a value for [ion]. No more than one significant figure should be used to express the final results. The method of calculating solubility product values from the results obtained is given in the *Students' book*.

Discussion

The results obtained by the whole class should be collected together and used to discuss briefly the relative solubility of the silver compounds studied. The values of the solubility products could be compared with those given in the *Book of data*.

Note. For the information of the teacher, it is not possible in practice to realize the extended graph of E against $\ln[Ag^+(aq)]$ used in Experiment 15.3b. If the dilution of silver nitrate solution is continued beyond the limit suggested in the experiment (10^{-4}M) the cell e.m.f. becomes constant in the way shown by the following results (obtained using a potentiometer):

$[Ag^+(aq)]/\text{mol dm}^{-3}$	10^{-1}	10^{-2}	10^{-3}	10^{-4}	10^{-5}	10^{-6}	10^{-7}	10^{-8}	10^{-9}
E/volt		0.41	0.37	0.31	0.25	0.20	0.19	0.17	0.18	0.17

It seems likely that the limiting value of E in this system is determined by the equilibrium

$$Ag^+(aq) + OH^-(aq) \rightleftharpoons AgOH(s)$$

since, when distilled water was used instead of $AgNO_3(aq)$ in the silver electrode system, the measured cell e.m.f. was 0.17 V, corresponding to a silver ion concentration of 3×10^{-6} mol dm^{-3}. This does not, of course, invalidate concentration measurements smaller than this value but in order to obtain lower values for $[Ag^+(aq)]$ a different system must be used. If $I^-(aq)$ ions are added to an electrode system which contains $OH^-(aq)$ ions and $Ag^+(aq)$ ions, the electrode potential of the system is determined by the equilibrium

$$Ag^+(aq) + I^-(aq) \rightleftharpoons AgI(s)$$

for which K_{sp} is much smaller than for the $Ag^+(aq)/OH^-(aq)$ equilibrium. A cell consisting of a standard copper electrode combined with a silver electrode dipping into potassium iodide solution gives a negative e.m.f. value considered from the cell diagram

$$Cu(s)|Cu^{2+}(aq)\vdots Ag^+(aq)|Ag(s)$$

of the same order as that given by the mixture of KI(aq) and $AgNO_3$(aq) in Experiment 15.3b. Some results for a range of electrolytes in the negative electrode are shown in table 15.1 opposite.

Background reading

At the end of this section of the *Students' book* there is a piece of Background reading on the pH meter. When this instrument was used, in Topic 12, it was not possible to explain how it works, because the explanation depends upon a knowledge of electrode potentials. However, it should now be possible for students to understand this interesting application of the variation of electrode

Electrolyte in Ag electrode (ratios by volume)	E/volt	ln [Ag^+(aq)] (graph)	[Ag^+(aq)] /mol dm^{-3}	[I^-(aq)] /mol dm^{-3}	K_{sp} /mol^2 dm^{-6}
0.1M Ag^+(aq)/ 0.1M I^-(aq) = 2:3	−0.36	−34.1	1.6×10^{-15}	2×10^{-2}	3×10^{-17}
0.1M Ag^+(aq)/ 0.1M I^-(aq) = 1:2	−0.37	−34.5	10^{-15}	3.3×10^{-2}	3×10^{-17}
0.1M Ag^+(aq)/ 0.1M I^-(aq) = 3:7	−0.38	−35.0	6.3×10^{-16}	4×10^{-2}	3×10^{-17}
0.1M I^-(aq)	−0.41	−36.4	1.6×10^{-16}	10^{-1}	2×10^{-17}

Table 15.1

potential with concentration. Because natural logarithms, ln, have been used almost exclusively in this course, it would be wise to remind students that pH numbers refer to logarithms *to base 10* of the hydrogen ion concentration, lg [H^+(aq)].

Suggestions for homework

Answering questions 8, 9, and 10 at the end of this Topic in the *Students' book*. Reading the Background reading on pH meters.

Summary

At the end of this section students should:

  1 know that concentration changes affect the electrode potentials of metal/metal ion systems;

  2 understand what is meant by a standard electrode potential, and be able to use these values to calculate the standard e.m.f. of cells;

  3 be able to use the Nernst equation;

  4 understand how to use electrode potentials to determine the solubilities of sparingly soluble salts.

15.4
REDOX EQUILIBRIA EXTENDED TO OTHER SYSTEMS
Objectives

  1 To study equilibria involving ion/ion and non-metal/non-metal ion reactions, and to use these in voltaic cells.

  2 To investigate the effect of concentration changes on E values for electrode systems based on ion/ion reactions.

3 To use $E^\ominus$ values to predict the possible course of redox reactions.
4 To use $E^\ominus$ values to find the equilibrium constants of redox reactions.

Timing
About one week.

Suggested treatment

For this treatment overhead projection transparency number 115 will be useful.
So far in this topic we have been concerned with redox systems involving a metal in equilibrium with its ions in solution. Experiment 15.4a introduces other types of redox reaction by a brief qualitative study of the reaction

$$2Fe^{3+}(aq) + 2I^-(aq) \longrightarrow 2Fe^{2+}(aq) + I_2(aq)$$

EXPERIMENT 15.4a
To investigate the reaction between iron(III) ions and iodide ions

Each student will need:
Test-tubes and rack

Access to solutions of the following materials, of approximately the concentrations indicated:
0.1M Fe^{3+}(aq); iron(III) sulphate or ammonium iron(III) sulphate (iron alum) are suitable
0.1M potassium iodide
0.1M Fe^{2+}(aq); iron(II) sulphate is suitable
1 per cent starch
2 per cent potassium hexacyanoferrate(III), $K_3Fe(CN)_6$

Procedure

Details of the procedure are given in the *Students' book*. Formation of a deep blue colour with potassium hexacyanoferrate(III) solution and with starch solution indicates that iron(II) and iodine are formed when solutions of iron(III) and iodide ions are mixed.

Use of the iron(III)/iodide reaction in a voltaic cell. The electron transfer which occurs during this reaction can be seen from the half-reactions

$$Fe^{3+}(aq) + e^- \longrightarrow Fe^{2+}(aq) \quad \text{reduction}$$
and $$2I^-(aq) \longrightarrow I_2(aq) + 2e^- \quad \text{oxidation}$$

which arise from the competing equilibria

$$Fe^{3+}(aq) + e^- \rightleftharpoons Fe^{2+}(aq)$$
and $$2I^-(aq) \rightleftharpoons I_2(aq) + 2e^-$$

The qualitative study enables the equilibrium position for the complete reaction

$$2Fe^{3+}(aq) + 2I^-(aq) \rightleftharpoons 2Fe^{2+}(aq) + I_2(aq)$$

to be established as lying towards the righthand side of the equation.

The use of this reaction in a voltaic cell is studied in Experiment 15.4b.

EXPERIMENT 15.4b
To measure the electrode potentials for the $Fe^{3+}(aq)/Fe^{2+}(aq)$ equilibrium and the $2I^{1-}(aq)/I_2(aq)$ equilibrium

Each pair of students will need:
Copper reference electrode, as in Experiment 15.2c
Potassium nitrate salt bridges, as in Experiment 15.2c   } or alternative versions suggested in Experiment 15.2c
2 smooth platinum electrodes
3 beakers, 50 cm³ or 100 cm³
Transistor or solid-state voltmeter

Access to:
Arbitrary solutions of:
   iodine in aqueous KI
   a mixture of iron(II) and iron(III) salts

The circuit is shown in figure 15.5 overleaf.

Procedure

Students measure the e.m.f.s of the cells:

$$Cu(s)|Cu^{2+}(aq) \vdots Fe^{3+}(aq), Fe^{2+}(aq)|Pt$$
and $$Cu(s)|Cu^{2+}(aq) \vdots I_2(aq), 2I^-(aq)|Pt$$

and from the results calculate the e.m.f. of the cell

$$Pt|2I^-(aq), I_2(aq) \vdots Fe^{3+}(aq), Fe^{2+}(aq)|Pt$$

They check their calculation by measuring the e.m.f. of this cell. A fresh potassium nitrate salt bridge should be used for each measurement.

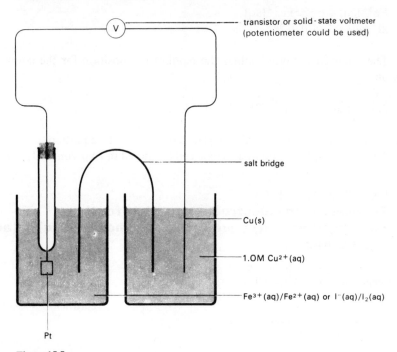

Figure 15.5

Discussion

Students' attention should be drawn to the fact that it is necessary to have both the oxidized and reduced forms in each electrode system, to enable the half-cell reactions to proceed in either direction.

The platinum electrode takes no part in the reactions but merely acts as an inert surface by means of which electrons can be transferred into or out of a half-cell.

The agreement between predicted and measured values for the third cell studied shows that ion/ion reactions and non-metal/non-metal ion reactions can be dealt with in the same way as metal/metal ion reactions studied earlier.

To deal with cells of this kind an extension of the conventions for writing cell diagrams is necessary. The reduced form of the redox couple is always placed nearest to the inert electrode and separated from the oxidized form by a comma

$$Fe^{3+}(aq), \quad Fe^{2+}(aq) | Pt$$
$$\text{oxidized} \quad \text{reduced}$$
$$\text{form} \quad \text{form}$$

Concentration effects in ion/ion systems

The application of either the principle of Le Châtelier or the equilibrium law to the iron(II)/iron(III) equilibrium indicates that the equilibrium position is affected by relative ion concentrations:

$$Fe^{3+}(aq) + e^- \rightleftharpoons Fe^{2+}(aq)$$

Increase in the relative concentration of $Fe^{3+}(aq)$ drives the equilibrium to the right, reducing the negative potential of the system and making the e.m.f. of the cell more positive.

Temperature also has an effect on the system so that for electrode systems of this kind it is necessary to specify concentration ratios and temperature for the standard redox potential ($E^\ominus$). The conditions chosen are:

Equal molar concentrations of reduced and oxidized forms.
A temperature of 298 K (25 °C).

The value of E for other conditions is given by the Nernst equation in the form

$$E = E^\ominus + \frac{RT}{zF} \ln \frac{[\text{oxidized form}]}{[\text{reduced form}]}$$

which gives for a temperature of 25 °C,

$$E = E^\ominus + \frac{0.026}{z} \ln \frac{[\text{oxidized form}]}{[\text{reduced form}]}$$

z is the number of electrons transferred when the oxidized form changes to the reduced form; for example, $z = 1$ for $Fe^{3+}(aq) + e^- \longrightarrow Fe^{2+}(aq)$.

Thus, using the values given in table 15.2 (which also appears in the *Students' book*) a plot of

$$\ln \frac{[Fe^{3+}(aq)]}{[Fe^{2+}(aq)]}$$

against E will give a straight line of slope approximately 0.026.

Relative concentrations/mol dm^{-3}		$\ln \dfrac{[Fe^{3+}]}{[Fe^{2+}]}$	E/volt
$[Fe^{3+}(aq)]$	$[Fe^{2+}(aq)]$		
1	9	-2.197	0.716
2	8	-1.386	0.735
3	7	-0.847	0.748
4	6	-0.405	0.760
5	5	0	0.770
6	4	$+0.405$	0.782
7	3	$+0.847$	0.792
8	2	$+1.386$	0.805
9	1	$+2.197$	0.825

Table 15.2
The variation of electrode potential with concentration for the $Fe^{3+}(aq)$, $Fe^{2+}(aq)$ electrode (E values measured against a standard hydrogen electrode).

When [oxidized form] = [reduced form]

$$\ln \frac{[\text{oxidized form}]}{[\text{reduced form}]} = \ln 1 = 0$$

and $E = E^{\ominus}$. This enables a value for $E^{\ominus}$ to be obtained for the iron(III)/iron(II) electrode from the graph by taking the value of E for $\ln[Fe^{3+}(aq)]/[Fe^{2+}(aq)] = 0$. It should be pointed out that the full Nernst equation

$$E = E^{\ominus} + \frac{RT}{zF} \ln \frac{[\text{oxidized form}]}{[\text{reduced form}]}$$

applies to all equilibria of this kind. In the case of a metal/metal ion system the reduced form is a crystalline metal; the concentration of this is constant so that variations in E depend on [ion] only.

As in preceding sections, attempts to deal more rigorously with the systems under consideration would involve the use of activities. The effects of inter-ionic attraction and incomplete ionization can be mentioned, but no more than this should be attempted.

Some uses of $E^{\ominus}$ values

The *Students' book* describes in some detail three uses for $E^{\ominus}$ values. They are:

1 *Calculating the e.m.f. of voltaic cells.* Students should be given some practice in this, using some of the questions at the end of this Topic in the *Students' book*.

2 *Predicting whether a reaction is likely to take place.* This is an extremely important subsection which should be gone through with care. It is important that the students can apply the 'anti-clockwise rule' correctly; they will be using it in Topic 16.

3 *Finding the equilibrium constant for a redox reaction.* Again an important subsection which should be given ample time. In this description, the Nernst equation for a metal/metal ion half cell is used to deduce the more general expression for a cell and then note to what this expression reduces when the cell is short-circuited and 'runs down' to equilibrium. There is a fairly full treatment of this in the *Students' book*, ending with a quantitative relationship between $E^{\ominus}_{\text{cell}}$ and K_c, namely

$$E^{\ominus}_{\text{cell}} = \frac{RT}{zF} \ln K_c.$$

In this expression the units of R are $J K^{-1} mol^{-1}$; R is therefore $8.31 J K^{-1} mol^{-1}$. With $F = 96\,500\,C$, this gives $E^{\ominus}$ in volts.

Table 15.3 in the *Students' book* lists some equilibrium constants for a selection of cell reactions. These K_c values have been calculated from the corresponding $E^{\ominus}_{\text{cell}}$ values using the Nernst equation.

Questions in the *Students' book*

The questions which are asked at this point in the *Students' book* are intended to enable students to familiarize themselves with the use of this equation. The answers are

1 Cell equilibrium reaction:

$$Cu(s) + Br_2(aq) \rightleftharpoons Cu^{2+}(aq) + 2Br^-(aq)$$

2 $K_c = 2 \times 10^{25}\,mol^2\,dm^{-6}$.

Summary

At the end of this section students should:

1 know that electrode potentials can be measured for ion/ion and non-metal/non-metal ion systems;

2 know how concentration changes affect E values in these other redox systems;

3 be able to use the anti-clockwise rule to predict the likely course of redox reactions, given a table of $E^{\ominus}$ values;

4 be able to calculate equilibrium constants for redox reactions from a knowledge of $E^{\ominus}$ values.

15.5
ENTROPY CONSIDERATIONS

Objectives

1 To develop a simplified picture of the entropy changes which result when an atom in a solid metal changes into an ion in solution during the working of a voltaic cell, based on previous treatment of similar changes when gases expand or contract.

2 To introduce the idea that the entropy increase which occurs when a metal atom changes into an ion varies with the concentration of the surrounding solution, being relatively larger when the solution is dilute than when it is concentrated.

3 To contrast the entropy changes in cells where the reactions taking place result in an overall change in the number of ions present with those in which the overall number of ions remains constant, using as examples cells in which the reactions are

$$2Ag^+(aq) + Cu(s) \longrightarrow 2Ag(s) + Cu^{2+}(aq)$$

(Loss of $Ag^+(aq)$ ions is greater than gain of $Cu^{2+}(aq)$ ions.)

and $\quad Cu^{2+}(aq) + Zn(s) \longrightarrow Zn^{2+}(aq) + Cu(s)$

(Loss of $Cu^{2+}(aq)$ ions equals gain of $Zn^{2+}(aq)$ ions.)

4 To examine the entropy changes occurring in system and surroundings when the copper/silver cell is working and to relate these to the maximum work that is theoretically obtainable from the cell ($W_{max} = -\Delta G$).

5 To relate free energy change to electromotive force in a cell, using the Daniell cell as an example, and to develop the relationship $\Delta G = -zFE_{cell}$.

6 To examine the effect of changing concentrations of electrode solutions on e.m.f. in the Daniell cell, and to explain this qualitatively by considering the entropy changes which take place at the electrodes.

Timing

No experimental work is included in this section, so four periods should be sufficient.

Suggested treatment

The *Students' book* sequence, for the most part, is qualitative in nature, and makes little demand on mathematical ability. It contains all that is required for examination purposes and some teachers, having in mind the capabilities of a given class, will want to follow it as it stands. One point might be stressed here. The explanation of the use of the adjective 'free' in 'free energy' (*Students' book* page 198) begins by referring to a system in which ΔH is more negative than ΔG. An example is the Ag/Cu cell, for which $\Delta G^\ominus = -89 \text{ kJ mol}^{-1}$ and $\Delta H^\ominus = -147 \text{ kJ mol}^{-1}$. Using the expression:

$$\Delta G^\ominus = \Delta H^\ominus - T\Delta S^\ominus$$

we have $T\Delta S^\ominus = \Delta H^\ominus - \Delta G^\ominus$

which for this cell

$$= -147 + 89$$
$$= -58 \text{ kJ mol}^{-1}$$

This figure is the energy that *must* be transferred to the surroundings. The case in which ΔG is more negative than ΔH should then be dealt with. An example is the Pb/Cu cell, with the cell reaction

$$\text{Pb(s)} + \text{Cu}^{2+}\text{(aq)} \longrightarrow \text{Pb}^{2+}\text{(aq)} + \text{Cu(s)}$$

for which $\Delta G^\ominus = -89 \text{ kJ mol}^{-1}$ and $\Delta H^\ominus = -63 \text{ kJ mol}^{-1}$.

In this case $T\Delta S^\ominus = -63 + 89$
$$= +26 \text{ kJ mol}^{-1}$$

which means that, for maximum work to be achieved, $+26 \text{ kJ mol}^{-1}$ must be transferred from the surroundings to the cell, that is, the cell must 'run cold'. This second example may be more difficult to understand, and may therefore require careful explanation.

Supporting homework

Answering questions at the end of this Topic in the *Students' book*.

Summary

At the end of this section students should:

1 appreciate that there is always an entropy increase when metal atoms go into solution as metal ions, but that this increase becomes smaller as the concentration of the solution increases;

2 appreciate the difference between standard and non-standard conditions when applied to discussions of voltaic cells;

3 be able to calculate the maximum work obtainable from a given voltaic cell and, hence, the value of $\Delta G^\ominus$ for the cell reaction;

4 be able to account qualitatively for the change in e.m.f. of a voltaic cell when the concentrations of the electrode solutions vary;

5 be able to use tables of appropriate data to calculate values of $\Delta H^\ominus$, $\Delta G^\ominus$, and $\Delta S^\ominus$ for given reactions.

15.6 PREDICTING WHETHER REACTIONS WILL TAKE PLACE; ΔG, E_{cell}, AND K_c

Objectives

To show how values of ΔG, E_{cell}, and K_c give information about whether or not a given change is feasible, under standard conditions, and how they can be used to indicate how far the change might proceed.

Timing

One period.

Suggested treatment

This section draws together threads which have been running through the course. Teachers should make sure that students understand its importance as a summary, and also appreciate the limitations of any conclusions drawn from it about reaction possibilities. It could be linked profitably with Topic 14, section 14.3, which deals with activation energies.

Supporting homework

Reading the section, before or after class discussion. Answering question 17 at the end of this Topic in the *Students' book*.

Summary

At the end of this section, students should be able to comment on the possibilities of reactions occurring, given values of $\Delta G^\ominus$, $E^\ominus_{cell}$, or K_c for the changes involved.

15.7
STANDARD FREE ENERGIES OF FORMATION, AND THEIR USE

Objectives

To define standard free energy of formation of a compound, and to show how tabulated values of these quantities can be used to calculate the standard free energy changes for reactions.

Timing

One or two periods should be sufficient for this section.

Suggested treatment

The treatment can be exactly parallel to that in Topic 6 (Energy changes and bonding) where the standard enthalpy change of formation of compounds was introduced. This is in fact a good point for students to revise that part of Topic 6.

The following points (which are treated more fully in the *Students' book*) could be brought out during the discussion.

1 Just as in the case of enthalpies, or of potential energy, absolute values of free energies are not known, so it is convenient to choose some base-line, or arbitrary zero, from which to measure the standard free energies of substances. The convention chosen for free energies is the same as that for enthalpies, namely that

    at 1 atm (101 325 Pa) pressure
    and 298 K
    with the elements in the physical states normal under these conditions,

the standard free energies of the elements are zero.

2 The standard free energy of formation of any compound at 298 K, $\Delta G^\ominus_{f,298}$ is the standard free energy change when one mole of the compound at 298 K is formed from its elements in physical states normal at 1 atm and 298 K.

124 Topic 15 Redox equilibria and free energy

3 It follows necessarily from this convention that $\Delta G^\ominus_{f,298}$ [elements in physical states normal at 1 atm and 298 K] = 0.

4 It is possible therefore, using this convention, to tabulate standard free energies of formation of *compounds* rather than standard free energies relating to specific reactions. (In fact, of course, the standard free energy of a compound does really refer to a reaction, namely the formation of one mole of a compound from its elements in physical states normal at 1 atm and 298 K.) The tabulation of standard free energy data relating to compounds rather than reactions, however, makes for very general and flexible use of such tables.

In dealing with ions, the standard free energy of the hydrogen ion is taken as zero, that is

$$\Delta G^\ominus_{f,298}[H^+(aq)] = 0$$

This corresponds to the convention of regarding the standard potential of the hydrogen electrode as zero volts.

How to calculate equilibrium constants from standard free energies of formation

Examples 1 to 3 in the *Students' book* are intended to show how the equilibrium constant for a reaction can be calculated from tabulated values of standard free energies of formation. The examples chosen are for reactions involving gases and reactions between ions in solution. The teacher can invent many other examples using the data given in the *Book of data*.

Further suggestions for teachers who wish to take the treatment of entropy further

Almost all the numerical data used in the treatment of energy relationships in this course have been those relating to *standard* conditions, which have been arbitrarily set at a temperature of 298 K, a pressure of 1 atmosphere, and, where solutions are involved, a concentration of 1 mol dm^{-3} of the reaction species (assuming no solvent/solute interactions). It is most important that students appreciate these limitations since the results obtained from calculations using $\Delta H^\ominus$, $\Delta G^\ominus$, and $S^\ominus$ values are often not applicable to other conditions. Although values of ΔH often do not vary very much with temperature and pressure for many substances, those for ΔG often do so quite widely. This difference is shown in table 15.3, which gives the variations of ΔH and ΔG for three reactions.

15.7 Standard free energies of formation, and their use

Reaction	T/K	ΔH_T/kJ	ΔG_T/kJ
$CaCO_3(s) \longrightarrow CaO(s) + CO_2(g)$	298	178	130
	500	177	98
	900	177	34
	1300	176	−30
$N_2(g) + 3H_2(g) \longrightarrow 2NH_3(g)$	298	−92	−33
	500	−101	14
	700	−110	55
	900	−119	103
$H_2O(g) + C(s) \longrightarrow H_2(g) + CO(g)$	298	131	91
	500	134	63
	900	141	4
	1300	147	−58

Table 15.3

In Topic 12 the relationship

$$S = S^\ominus - Lk \ln p/p^\ominus$$

was used to deal with the effects of a pressure change from $p^\ominus$ to p in gaseous reactions. Figure 15.6 shows, in diagrammatic form, that the molecules of a gas have more ways of occupying the increased volume which results from a lowering of pressure, and hence the entropy rises.

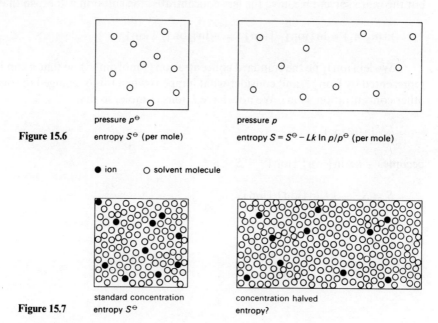

Figure 15.6 pressure $p^\ominus$ / entropy $S^\ominus$ (per mole)   pressure p / entropy $S = S^\ominus - Lk \ln p/p^\ominus$ (per mole)

● ion   ○ solvent molecule

Figure 15.7 standard concentration / entropy $S^\ominus$ concentration halved / entropy?

There are evidently similarities between gases and dilute solutions of electrolytes, which might be represented as in figure 15.7. Instead of having molecules occupying spaces, as with gases, we now have ions occupying places; not, of course, in fixed positions but momentarily taking up places that would otherwise belong to solvent molecules. If the concentration of the solution is halved the number of solvent molecules is doubled for a *fixed number of ions*, and each ion has twice as many choices of where to go. Provided that the solution is fairly dilute, the choice for one ion can be independent of that for another ion. Hence if there are N ions altogether, doubling the number of solvent molecules multiplies the number of ways by 2^N. More generally, just as with gases expanding from V_1 to V_2, changing from n_1 to n_2 solvent molecules multiplies the number of ways for N ions by $(n_2/n_1)^N$. Since for any change of this kind in the number of ways

$$\Delta S = k \Delta \ln W \quad (k \text{ is the Boltzmann constant}, 1.38 \times 10^{-23} \text{ J K}^{-1})$$

the dilution of the solution from n_1 solvent molecules to n_2 solvent molecules results in an entropy change

$$\Delta S = k \ln (n_2/n_1)^N$$
$$= kN \ln (n_2/n_1)$$

but the more solvent we have the less concentrated the solution will be, so that

$$\ln(n_2/n_1) = \ln [\text{ion}]_1/[\text{ion}]_2 = -\ln [\text{ion}]_2/[\text{ion}]_1$$

We let $[\text{ion}]_1$ be the standard concentration (1 mol dm^{-3}), so that it can be represented by $[\text{ion}]^\ominus$, and consider what happens when this is changed to some other concentration, $[\text{ion}]$. We now have L ions (1 mole) so that

$$kN \ln (n_2/n_1)$$

becomes $-kL \ln [\text{ion}]/[\text{ion}]^\ominus = S - S^\ominus$

$$\therefore \quad S = S^\ominus - kL \ln [\text{ion}]/[\text{ion}]^\ominus$$
$$= S^\ominus - R \ln [\text{ion}]/[\text{ion}]^\ominus$$

but $[\text{ion}]^\ominus = 1$

$$\therefore \quad S = S^\ominus - R \ln [\text{ion}]$$

Per mole of ions a more dilute solution has a bigger entropy just because there are more solvent molecules with which ions can be interchanged.

The above equation is not accurate in all conditions. Obviously, it cannot work sensibly at high concentrations because the ratio of ions to molecules is high. In the absurd limit it would be 'all ions'. But in a molar aqueous solution there are nearly 60 water molecules to each ion. Interchanging one ion with another just like it makes no difference to the number of ways, so there must be few enough ions for this not to happen very often.

Nevertheless, the relationship

$$S = S^{\ominus} - R\ln[\text{ion}]$$

is often a fair sort of guide, and will generally predict correctly whether the entropy goes up or down, if not precisely by how much. No very accurate molecular picture of ionic solutions has yet been produced, so when the equation is not correct, it is normally patched up by an empirical correction. This correction uses 'activity coefficients', which are introduced to correct for the electrostatic attractive forces which operate between oppositely charged particles in electrolyte solutions. These forces have the effect of restricting the mobility of the individual particles and hence reduce the 'effectiveness' or 'activity' of the ionic species as far as their chemical behaviour is concerned. To allow for this departure from 'ideal' behaviour, activity coefficients are introduced, defined as the ratio of the effective concentration, or activity, to the stoicheiometric concentration of the species, that is,

$$f_X = \frac{a_X}{[X]} \quad \text{or} \quad a_X = [X]f_X$$

where a_X is the activity of X
 [X] is the stoicheiometric concentration of X
 f_x is the activity coefficient of X

The value of the activity coefficient depends on a number of factors, including ionic charge, ionic size, temperature, and total ionic concentration in solution. Table 15.4 may help to illustrate this.

Ions (in aqueous solution)	Ionic diameter/nm	Total ionic concentration/mol dm^{-3}			
		0.001	0.01	0.1	0.2
		Activity coefficient (f_X)			
H^+	0.9	0.975	0.933	0.860	0.83
K^+, Ag^+, Cl^-, Br^-	0.3	0.975	0.925	0.805	0.76
$Sr^{2+}, Ba^{2+}, SO_4^{2-}$	0.5	0.903	0.744	0.465	0.39
$Ca^{2+}, Fe^{2+}, Zn^{2+}$	0.6	0.905	0.749	0.485	0.42
Al^{3+}, Fe^{3+}	0.9	0.802	0.540	0.245	0.17

Table 15.4
Approximate individual activity coefficients of ions in water at 25 °C.

The use of activity coefficients may be illustrated from the equilibrium

$$Ag^+(aq) + Fe^{2+}(aq) \rightleftharpoons Fe^{3+}(aq) + Ag(s)$$

The value of K_c for this can be calculated using tables of standard electrode potentials.

for $Fe^{3+}(aq), Fe^{2+}(aq) | Pt$; $\quad E^\ominus = +0.77\,V$

and for $Ag^+(aq)|Ag(s)$; $\quad E^\ominus = +0.80\,V$

Hence for the cell

$$Pt|Fe^{2+}(aq), Fe^{3+}(aq) \vdots Ag^+(aq)|Ag(s)$$

$E^\ominus = \qquad -0.77 \qquad\qquad +0.80 \qquad = 0.03\,V$

$$E^\ominus = \frac{RT}{zF} \ln K_c$$

$$\therefore \quad 0.03 = \frac{8.31 \times 298}{1 \times 96\,500} \ln K_c$$

$$\therefore \quad \ln K_c = \frac{0.03 \times 96\,500}{8.31 \times 298} = 1.17$$

and $K_c = 3.2\,dm^3\,mol^{-1}$

This could be called the 'theoretical' value for K_c, since activity values are allowed for when calculating $E^\ominus$ values. K_c for this equilibrium can also be obtained using volumetric methods, which assume complete ionization of reagents and no 'congestion' factors; the value then obtained is notably higher than $3.2\,dm^3\,mol^{-1}$ and can be called the 'experimental' value.

To sum up

$$K_{c\text{ theoretical}} = \frac{a_{\text{Fe}^{3+}(\text{aq}) \text{ eqm}}}{a_{\text{Ag}^{+}(\text{aq}) \text{ eqm}} \times a_{\text{Fe}^{2+}(\text{aq}) \text{ eqm}}}$$

and $$K_{c\text{ experimental}} = \frac{[\text{Fe}^{3+}(\text{aq})]_{\text{eqm}}}{[\text{Ag}^{+}(\text{aq})]_{\text{eqm}}[\text{Fe}^{2+}(\text{aq})]_{\text{eqm}}}$$

Since $a_{\text{Fe}^{3+}(\text{aq})\text{eqm}} = [\text{Fe}^{3+}(\text{aq})]_{\text{eqm}} \times f_{\text{Fe}^{3+}(\text{aq})}$

$a_{\text{Ag}^{+}(\text{aq})\text{eqm}} = [\text{Ag}^{+}(\text{aq})]_{\text{eqm}} \times f_{\text{Ag}^{+}(\text{aq})}$

and $a_{\text{Fe}^{2+}(\text{aq})\text{eqm}} = [\text{Fe}^{2+}(\text{aq})]_{\text{eqm}} \times f_{\text{Fe}^{2+}(\text{aq})}$

$$K_{c\text{ theoretical}} = \frac{[\text{Fe}^{3+}(\text{aq})]_{\text{eqm}} \times f_{\text{Fe}^{3+}(\text{aq})}}{[\text{Ag}^{+}(\text{aq})]_{\text{eqm}} \times f_{\text{Ag}^{+}(\text{aq})} \times [\text{Fe}^{2+}(\text{aq})]_{\text{eqm}} \times f_{\text{Fe}^{2+}(\text{aq})}}$$

$$= K_{c\text{ experimental}} \times \frac{f_{\text{Fe}^{3+}(\text{aq})}}{f_{\text{Ag}^{+}(\text{aq})} \times f_{\text{Fe}^{2+}(\text{aq})}}$$

For a total ionic concentration of 0.2M using the values for activity coefficients in table 15.4, we have:

$$K_{c\text{ theoretical}} = K_{c\text{ experimental}} \times \frac{0.17}{0.76 \times 0.42}$$

$$= K_{c\text{ experimental}} \times 0.53$$

or $K_{c\text{ experimental}} \approx 2 \times K_{c\text{ theoretical}}$

So we should expect the value of K_c obtained by using standard electrode potentials (where allowance is made for incomplete ionization and for ion/ion and ion/solvent interactions) to be approximately half that obtained by volumetric methods. In fact values of $K_{c\text{ experimental}}$ greater than 6.4 dm^3 mol^{-1} are often obtained, probably owing to the difficulty in obtaining solutions of iron(II) salts which are free from iron(III) ions.

The Nernst equation

The relationship

$$S = S^{\ominus} - R \ln [\text{ion}]$$

is the basis of the Nernst equation, which is quoted in the *Students' book*, but no attempt is made to derive it. For able students this could be discussed on, of course, a strictly non-examinable basis.

In the *Students' book*, values are given for the e.m.f. of a Daniell cell under non-standard conditions of concentration, so it would be convenient to use these as a basis from which to proceed. In this cell the electrode reactions are

$$Zn(s) \longrightarrow Zn^{2+}(aq) + 2e^-$$
and $Cu^{2+}(aq) + 2e^- \longrightarrow Cu(s)$

If we concentrate initially on change in potential of the zinc electrode when $Zn^{2+}(aq)$ is altered*

$$S = S^\ominus - R \ln [Zn^{2+}]$$

and, therefore, for a change in S

$$\Delta S = \Delta S^\ominus - R \ln [Zn^{2+}]$$

The value of $E\{Zn^{2+}(aq)|Zn(s)\}$ refers to the process

$$Zn^{2+}(aq) + 2e^- \longrightarrow Zn(s)$$

Since what actually happens in the Daniell cell is the reverse of this

$$Zn(s) \longrightarrow Zn^{2+}(aq) + 2e^-$$

we must use $-E\{Zn^{2+}(aq)|Zn(s)\}$ and will replace this by E' for brevity.

E is related to ΔG by

$$\Delta G = -zFE$$

so that ΔG (for free energy change at zinc electrode) $= -(-zFE')$
$= zFE'$
but $\Delta G = \Delta H - T\Delta S$
$\therefore \quad zFE' = \Delta H - T\Delta S$

Under standard conditions, when $[Zn^{2+}(aq)] = 1.0M$

$$zFE^\ominus = \Delta H^\ominus - T\Delta S^\ominus$$

---

* Of course, division of the cell potential into separate electrodes is only a convenience to simplify discussion. Absolute values for single electrode potentials can never be obtained.

If the concentration of $Zn^{2+}(aq)$ is altered, say to 0.1M, ΔH will remain unchanged, but ΔS will increase, because of dilution, hence E' must change. The new value of ΔS is

$$\Delta S = \Delta S^{\ominus} - R \ln [Zn^{2+}(aq)]$$ and putting this in the equation

$$zFE' = \Delta H^{\ominus} - T\Delta S$$

gives $zFE' = \Delta H^{\ominus} - T(\Delta S^{\ominus} - R \ln [Zn^{2+}(aq)])$
$$= \Delta H^{\ominus} - T\Delta S^{\ominus} + RT \ln [Zn^{2+}(aq)]$$
but $\Delta H^{\ominus} - T\Delta S^{\ominus} = zFE^{\ominus}$
$$\therefore \quad zFE' = zFE^{\ominus} + RT \ln [Zn^{2+}(aq)]$$

dividing through by zF gives

$$E' = E^{\ominus} + \frac{RT}{zF} \ln [Zn^{2+}(aq)]$$

This expression makes it possible to calculate the electrode potential of the $Zn^{2+}(aq)|Zn$ electrode for any value of $[Zn^{2+}(aq)]$. It is the Nernst equation, which can be simplified to

$$E' = E^{\ominus} + \frac{0.026}{z} \ln [Zn^{2+}(aq)]$$

For $[Zn^{2+}(aq)] = 0.1 \, mol \, dm^{-3}$

$$E' = E^{\ominus} + \frac{0.026}{2} \ln 0.1$$
$$= E^{\ominus} - 0.03 \, V$$

$E^{\ominus}$ for the zinc electrode is $-0.76\,V$, so when $[Zn^{2+}(aq)] = 0.1$, $E = -0.79\,V$, which combined with a standard copper electrode ($E^{\ominus} = +0.34\,V$) gives a Daniell cell e.m.f. of 1.13 V. This explains the information given in the *Students' book* that E_{cell} increases by 0.03 V when $[Zn^{2+}(aq)]$ is reduced from 1.0 to $0.1 \, mol \, dm^{-3}$.

A similar argument can be used to explain why E_{cell} is *reduced* when the concentration of $Cu^{2+}(aq)$ is lowered. The situation here is exactly the reverse of the change in $[Zn^{2+}(aq)]$; $Cu^{2+}(aq)$ ions are being *removed* from solution, thus causing a reduction in entropy. The effect of changing $Cu^{2+}(aq)$ concentration is therefore the opposite of the zinc case.

Background reading

Finally in this section of the *Students' book* there is a piece of Background reading entitled 'Energetics in life processes'. This shows how the subject matter of this Topic is applied to related fields, in this case the subject of photosynthesis.

Summary

At the end of this section students should:
1 know what is meant by the standard free energy of formation of a compound;
2 be able to use tables of standard free energies of formation to calculate the standard free energy changes for reactions;
3 be aware of some aspects of the energetics of life processes.

ANSWERS TO PROBLEMS IN THE *STUDENTS' BOOK*

(A suggested mark allocation is given in brackets after each answer.)

1a Fe(s) reductant, Cu^{2+}(aq) oxidant (1)
 b Al(s) reductant, H^+(aq) oxidant (1)
 c Zn(s) reductant, Pb^{2+}(aq) oxidant (1)
 d Sn^{2+}(aq) reductant, Fe^{3+}(aq) oxidant (1)

 Total 4 marks

2a Metals. (1)
   All form positive ions in aqueous solution (2)
 b D(s) + $2C^+$(aq) ⟶ D^{2+}(aq) + 2C(s) (2)
 c i A(s) ⟶ A^{2+}(aq) + $2e^-$
     B^{2+}(aq) + $2e^-$ ⟶ B(s) (2)
   ii A(s) ⟶ A^{2+}(aq) + $2e^-$
     $2C^+$(aq) + $2e^-$ ⟶ 2C(s) (2)
   iii B(s) ⟶ B^{2+}(aq) + $2e^-$
     $2C^+$(aq) + $2e^-$ ⟶ 2C(s) (2)
   iv D(s) ⟶ D^{2+}(aq) + $2e^-$
     B^{2+}(aq) + $2e^-$ ⟶ B(s) (2)
 d A, D, B, C (2)
   A reduces B^{2+}(aq) and C^+(aq)
   D does not reduce A^{2+}(aq)
   D reduces B^{2+}(aq) and
   B reduces C^+(aq) (5)

   Total 20 marks

3a H_2 positive, Fe negative (1)
 $E^\ominus = -0.44$ volt (2)
b Ni negative, H_2 positive (1)
 $E^\ominus = +0.25$ volt (2)
c Zn negative, Ni positive (1)
 $E^\ominus = 0.76 - 0.25 = +0.51$ volt (2)
d Al negative, Cr positive (1)
 $E^\ominus = 1.66 - 0.74 = +0.92$ volt (2)
 Total 12 marks

4 Standard electrode potential $= 0.34 - 0.62$
 $= -0.28$ volt **Total 2 marks**

5 Standard electrode potential $= 1.61 - 0.76$
 $= +0.85$ volt **Total 2 marks**

6a $Al(s) \longrightarrow Al^{3+}(aq) + 3e^-$
 $Sn^{2+}(aq) + 2e^- \longrightarrow Sn(s)$
 $2Al(s) + 3Sn^{2+}(aq) \longrightarrow 3Sn(s) + 2Al^{3+}(aq)$ (3)
b $Pb(s) \longrightarrow Pb^{2+}(aq) + 2e^-$
 $Ag^+(aq) + e^- \longrightarrow Ag(s)$
 $Pb(s) + 2Ag^+(aq) \longrightarrow 2Ag(s) + Pb^{2+}(aq)$ (3)
c $Mg(s) \longrightarrow Mg^{2+}(aq) + 2e^-$
 $2H^+(aq) + 2e^- \longrightarrow H_2(g)$
 $Mg(s) + 2H^+(aq) \longrightarrow Mg^{2+}(aq) + H_2(g)$ (3)
 Total 9 marks

7a $Ag^+(aq)$ $Cu^{2+}(aq)$ $Pb^{2+}(aq)$ $Cr^{3+}(aq)$ (2)
b $Fe^{3+}(aq)$ $Sn^{2+}(aq)$ $Zn^{2+}(aq)$ $Mg^{2+}(aq)$ (2)
 Total 4 marks

8a The graph of electrode potential against ln [ion] is as shown in figure 15.8. (5)
b -0.277 volt (2)
 $E = E^\ominus$ when ln [ion] $= 0$ (3)
c Slope of graph $= 0.013$ (2)
 z (charge on ion) $= 2$ (2)
 As M is a metal, charge on ion $= M^{2+}$ (1)
 Total 15 marks

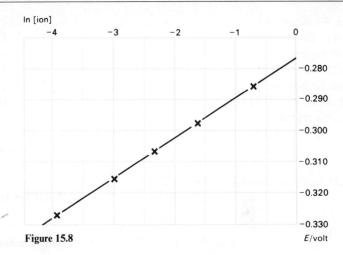

Figure 15.8

9a From the Nernst equation
$\ln[Ag^+(aq)] = -7.3$
$[Ag^+(aq)] = 6.7 \times 10^{-4}\,\text{mol dm}^{-3}$
But marks should be given for correct reading from the student's own graph of *electrode potential* against $\ln[Ag^+(aq)]$ obtained from Experiment 15.3a (2)

b $K_{sp} = [Ag^+(aq)]_{eqm}[BrO_3^-(aq)]_{eqm}$ (2)
$10\,\text{cm}^3$ of the $0.1\,\text{M}$ $KBrO_3(aq)$ precipitate
$\therefore 40\,\text{cm}^3$ remain in a total volume of $60\,\text{cm}^3$ (2)

$\therefore [BrO_3^-]$ in final solution $= \dfrac{40}{60} \times 10^{-1}\,\text{mol dm}^{-3}$ (2)

$\therefore K_{sp} = [Ag^+(aq)] \times \dfrac{1}{15}\,\text{mol}^2\,\text{dm}^{-6}$ (2)

Total 10 marks

10 The graph of $\ln\dfrac{[M^{x+}(aq)]}{[M^{2+}(aq)]}$ against E should look like figure 15.9. (5)
$E^\ominus = 0.14\,\text{volt}$ (2)
$E^\ominus = E$ when $\ln\dfrac{[M^{x+}]}{[M^{2+}]} = 0$ (3)
Slope of graph $= 0.013$ (2)
$\therefore z =$ number of electrons transferred when oxidized form
changes to reduced form $= 2$ (2)
$\therefore x = 4$ (1)

Total 15 marks

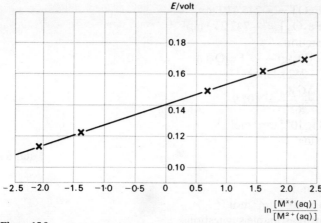

Figure 15.9

11a -65.0 kJ (2)
b -2.9 kJ mol^{-1} (3)
Total 5 marks

12a Pt|I$^-$(aq), I$_2$(aq)|Br$_2$(aq), Br$^-$(aq)|Pt (1)
b $+0.53$ V (1)
c 2×10^9 (for equation as written in the question) (3)
Total 5 marks

13a MnO$_4^-$(aq) + 8H$^+$(aq) + 5e$^- \rightleftharpoons$ Mn^{2+}(aq) + 4H$_2$O(l) (2)
 and 5Fe^{3+}(aq) + 5e$^- \rightleftharpoons$ 5Fe^{2+}(aq) (2)
b Pt|Fe^{2+}(aq), Fe^{3+}(aq)|[MnO$_4^-$(aq) + 8H$^+$(aq)],
 [Mn^{2+}(aq) + 4H$_2$O(l)]|Pt (2)
c $+0.74$ V (2)
d $K_c = 4 \times 10^{62}$ (2)
Total 10 marks

14a -50.8 kJ (1)
b Yes (1)
c No (1)
d Answer should refer to distinction between kinetic and energetic factors. (2)
Total 5 marks

15a $\Delta G^\ominus_{f,298}[\text{FeO}] = -245.4\,\text{kJ}\,\text{mol}^{-1}$
$\Delta G^\ominus_{f,298}[\text{Fe}_2\text{O}_3] = -742.2\,\text{kJ}\,\text{mol}^{-1}$ (2)
b For the reaction
$2\text{FeO(s)} + \tfrac{1}{2}\text{O}_2(\text{g}) \longrightarrow \text{Fe}_2\text{O}_3(\text{s})$
$\Delta G^\ominus_{298} = -252\,\text{kJ}\,\text{mol}^{-1}$
$\ln K_c = -\dfrac{\Delta G^\ominus_{298}}{RT} = 101.9$
$K_c = 1.84 \times 10^{44}\,\text{dm}^3\,\text{mol}^{-1}$ (3)
c Iron(III) oxide (1)
Total 6 marks

16a $-147.5\,\text{kJ}\,\text{mol}^{-1}$ (2)
b $K_c = 7.38 \times 10^{25}\,\text{dm}^6\,\text{mol}^{-2}$ (2)
c 4 marks for reasonable suggestions. (4)
Total 8 marks

17a **i** $\Delta G^\ominus = -394.4\,\text{kJ}\,\text{mol}^{-1}; \Delta H^\ominus = -393.5\,\text{kJ}\,\text{mol}^{-1}$. (4)
  ii $\Delta G^\ominus = -33.0\,\text{kJ}\,\text{mol}^{-1}; \Delta H^\ominus = -92.2\,\text{kJ}\,\text{mol}^{-1}$. (4)
  iii $\Delta G^\ominus = +130.4\,\text{kJ}\,\text{mol}^{-1}; \Delta H^\ominus = +178.3\,\text{kJ}\,\text{mol}^{-1}$. (4)
  iv $\Delta G^\ominus = -212.5\,\text{kJ}\,\text{mol}^{-1}; \Delta H^\ominus = -218.7\,\text{kJ}\,\text{mol}^{-1}$. (4)
b Reactions **i** and **iv** above agree closely; the others differ significantly. (2)
c ΔG and ΔH are most likely to show close agreement in reactions that involve the same numbers of particles on both sides of the equation, and have reactants and products in the same state. (4)
d ΔH is a good guide to the feasibility of a reaction because entropy changes of the surroundings (measured by $\Delta H/T$) at room temperature are usually much larger than those of the system, and thus usually outweigh them. ΔG is a better guide because it takes account of both entropy changes. (4)
Total 26 marks

18a $\text{Ni(s)}\,|\,\text{Ni}^{2+}(\text{aq})\,\vdots\,\text{Cu}^{2+}(\text{aq})\,|\,\text{Cu(s)}; E^\ominus = +0.59\,\text{V}$. (4)
b $\text{Ni(s)} + \text{Cu}^{2+}(\text{aq}) \longrightarrow \text{Ni}^{2+}(\text{aq}) + \text{Cu(s)}$. (2)
c Using $\Delta G^\ominus = -zFE^\ominus$
$\Delta G^\ominus = -2 \times 96\,500 \times 0.59 = -113\,870\,\text{J}\,\text{mol}^{-1}$ or $-113.87\,\text{kJ}\,\text{mol}^{-1}$. (2)
d $113.87\,\text{kJ}\,\text{mol}^{-1}$. (2)
e The maximum amount of work possible is calculated from the e.m.f. of the cell, obtained when no current is flowing. In order to obtain work from the cell current must flow, and when this happens the potential difference across the electrodes falls. The work that is obtained depends upon this potential difference and is thus less than the maximum. (6)
Total 16 marks

TOPIC 16
The Periodic Table 4: the transition elements

OBJECTIVES

1 To draw attention to some of the special characteristics of the transition elements.
2 To carry out investigations on various transition elements to illustrate these special properties.
3 To provide opportunities to use standard electrode potentials to predict reactions.
4 To provide illustrations of the application of thermodynamics to chemical systems.

CONTENT

16.1 The special properties of the transition elements. Electronic structure; similarity of physical properties; variable oxidation number; formation of complex ions; colour; catalytic activity.
16.2 Variable oxidation number. Experimental investigation of redox reactions of iron and of vanadium; analysis of 'iron tablets'.
16.3 Complex ion formation. Stability constants; experimental investigation of copper(II) complexes: stoicheiometry of complexes and method of investigation; preparation of salts containing complex ions.
16.4 Entropy considerations. Changes in numbers of particles involved and their effect on entropy changes.
16.5 Transition elements as catalysts. Homogeneous and heterogeneous catalysis; adsorption in heterogeneous catalysis; choosing a catalyst for the synthesis of ammonia; standard electrode potentials and homogeneous catalysts. Background reading: 1 'Ellingham diagrams and the iron and steel industry'; 2 'Micronutrients'.

TIMING

About two and a half weeks will be required for this Topic.

INTRODUCTION

A transition element is defined as one which contains an incomplete d-shell in at least one *compound*. It is not synonymous with a d-block element. The distinction between these two terms is made clear at the start of this Topic in the *Students' book*. The effect of the distinction is that scandium and zinc, two d-block elements which do not have the typical properties of transition elements, are not classified as such. Only the first row of transition elements are considered in this Topic, that is, the elements titanium to copper inclusive.

16.1
THE SPECIAL PROPERTIES OF THE TRANSITION ELEMENTS

Objectives

1 To define transition elements in terms of electronic structure.
2 To outline the characteristic properties of the transition elements.

Timing

Two periods.

Suggested treatment

For this treatment the teacher will find the following items helpful.
Samples of transition elements and some of their alloys.
Samples of a selection of compounds of the d-block elements in which these elements have as many different oxidation numbers as possible.
Overhead projection transparency numbers 116 to 118.

It will also be useful to have an energy level display board and an oxidation number display board (see Appendix 3).
An introduction is given in the *Students' book*, and students are then expected to write their own account of this selection of special properties, expanding it by reference to the *Book of data* and textbooks of inorganic chemistry. The teacher can help in this process by:
1 drawing attention to sources of information;
2 pointing out and discussing the anomalies in electronic structure;
3 explaining what a complex is;
4 discussing variable oxidation number.

A useful reference is GADD, K. F. 'Ionization energies and d-orbitals' *School Science Review*, **212**, 68, 526 (1979).

To introduce this topic, the following points should be covered.

16.1 The special properties of the transition elements

1 The definition of a *transition element*, and how it differs from a *d-block element* (see introduction to this topic).

2 All these elements have many similar properties.

a All are metals, and possess many similarities in their physical properties.

b All exhibit variable oxidation number. A discussion of this property is set out in section 16.2.

c All form many complex ions. These are discussed in section 16.3.

d All have coloured ions, that is, their ions absorb light in the visible part of the spectrum.

e All exhibit catalytic activity. This is discussed in section 16.5.

Observation of samples of the elements, their compounds, and the Periodic Table will bring out these points.

In Topic 4 it was found that the 2,3,3 grouping of first ionization energies was broken after calcium, and that the ten d-block elements produced this break. Furthermore, a new energy level, the d level, belonging to the $n = 3$ quantum level, became occupied after calcium. This level comes just above the 4s level and below the 4p level, as indicated in figure 16.1.

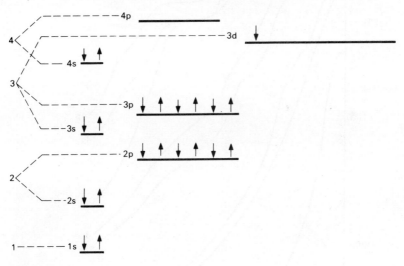

Figure 16.1
Electronic structure of scandium.

As the atomic number increases from 21 to 30 each element has one more electron in the 3d level than the previous elements, until the d level is full (10 electrons); then gallium has a full 3d level and one electron in the 4p level.

The teacher may care to note that the electrons do not fill the 3d level in a straightforward manner, as in some cases an arrangement with only one electron in the 4s level and an extra one in the 3d level may be more stable. In fact the 3d level is below the 4s from scandium onwards but the two levels are very close in energy. The teacher should use as much of this information as seems appropriate. All students should realize that the 3d and 4s levels are very close and that therefore the most stable arrangement of electrons may not necessarily have two electrons in the 4s level. The change of energy level with atomic number is shown in figure 16.2.

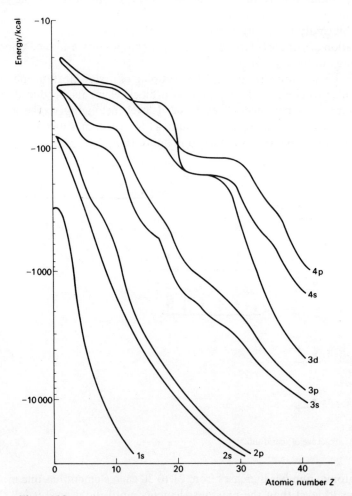

Figure 16.2
Change of energy level with atomic number (1 kcal = 4.2 kJ). *After Harvey, K. B. and Porter, G. B. (1963)* Introduction to physical inorganic chemistry, *Addison Wesley.*

Summary

Students should now know:
1 what is meant by the term transition element;
2 the characteristic features of transition elements.

16.2
VARIABLE OXIDATION NUMBER
Objectives

1 To investigate experimentally the reactions of iron compounds in which iron has different oxidation numbers, and to compare these with predictions based on standard electrode potentials.
2 To do the same for vanadium.
3 To show how to estimate the percentage of iron in an iron compound.

Timing

Six periods will be needed for the experiments, and some time for discussion.

Suggested treatment

For this treatment overhead projection transparencies numbers 116 and 117 will be helpful.

In the first experiment students investigate the reactions of Fe^{2+} and Fe^{3+} compounds. The *Students' book* gives the standard electrode potentials needed to predict the outcome of the reactions, and full details of the experiment.

EXPERIMENT 16.2a
An investigation of the redox reactions of iron

Each student, or pair of students, will need:
Rack of test-tubes
Dropping pipette
Access to zinc powder
0.1M $FeSO_4$ in dilute sulphuric acid, about 40 cm^3
0.1M $FeCl_3$ in dilute hydrochloric acid, about 40 cm^3
0.1M bromine solution, 10 cm^3
0.1M potassium manganate(VII), 10 cm^3
0.1M chlorine solution, 10 cm^3
0.1M sodium chloride, 10 cm^3
Sulphur dioxide solution, 10 cm^3 (CARE: may be harmful to those who suffer from respiratory complaints)
Silver nitrate solution, 10 cm^3 (0.05M or less is suitable)
0.1M potassium iodide, 10 cm^3

Procedure

Full details of the procedure are given in the *Students' book*. The outcome of the experiments is as follows.

a Iron(II) is oxidized to iron(III) by bromine water; the colour of bromine disappears.

b Iron(III) is reduced to iron(II) by zinc; slight colour change; NaOH test gives green colour of iron(II) hydroxide.

c Iron(II) reduces Ag^+ to silver metal; grey precipitate.

d No reaction.

e Iron(III) is reduced to iron(II) by sulphur dioxide; the reaction is quite slow; NaOH test gives green colour of iron(II) hydroxide.

f Iron(II) is oxidized by potassium manganate(VII); purple colour disappears.

g Iron(III) oxidizes iodide ions to iodine; 'iodine colour' produced.

h Iron(II) is oxidized to iron(III) by chlorine water; NaOH test gives red-brown colour of iron(III) hydroxide.

Next, students investigate the reactions of vanadium. It is easy to obtain compounds having this element with any one of four different oxidation numbers.

EXPERIMENT 16.2b
An investigation of the redox reactions of vanadium

Each student, or pair of students, will need:
Rack of test-tubes

Access to:
0.1M ammonium vanadate(V) (polytrioxovanadate(V)) in dilute sulphuric acid, $20 \, cm^3$
Zinc powder
Granulated tin
Copper powder
0.1M potassium iodide, $10 \, cm^3$
0.1M $FeSO_4$ in dilute sulphuric acid, $10 \, cm^3$
0.1M $FeCl_3$ in dilute hydrochloric acid, $20 \, cm^3$
0.1M potassium bromide, $10 \, cm^3$
0.1M copper(II) sulphate, $10 \, cm^3$
0.5M sodium thiosulphate, $10 \, cm^3$

Procedure

Full details are given in the *Students' book*.
The outcome of the experiment is as follows.

Part 1

VO^{2+} can be obtained using potassium iodide solution followed by sodium thiosulphate, or by using tin and decanting the solution when the reaction has

reached the appropriate stage.
V^{3+} can be obtained using granulated tin, but the reaction is quite slow.
V^{2+} can be obtained using powdered zinc, but again the reaction is quite slow.

Part 2
VO^{2+} and V^{2+}: the green colour of V^{3+} should be seen.
VO_2^+ and V^{3+}: the blue colour of VO^{2+} should be seen but excess of either reagent will make the solution appear green.
VO^{2+} and Fe^{3+}: no reaction is expected.
VO^{2+} and Br^-: no reaction is expected.
V^{2+} and Cu^{2+}: a red precipitate of copper should be formed.
V^{3+} and Fe^{3+}: a blue solution containing VO^{2+} should be produced but excess of either reactant will make the solution appear green.

Further references for vanadium compounds:
REDSHAW, D. J. 'Investigation of the oxidation numbers of vanadium.' *School Science Review*. **193**, *55*, 753 (1974).
BRITTEN, G. C. ET AL. 'Vanadium chemistry: some small scale methods.' *School Science Review*. **204**, *58*, 486 (1977).

As an alternative experiment, the oxidation numbers of molybdenum have been described in MACPHERSON, J. 'Changing the oxidation states of molybdenum'. *School Science Review*, **202**, *58*, 82 (1976).

In Experiment 16.2c, students analyse 'iron tablets' and calculate the percentage of iron in the tablets.

EXPERIMENT 16.2c
Estimation of the percentage of iron in 'ferrous sulphate' tablets

Each student, or pair of students, will need:
Access to balance, reading to 0.001 g
Burette, 50 cm³
Funnel
Pipette, 10 cm³
Pipette filler
Volumetric flask, 100 cm³
Conical flask, 250 cm³
Pestle and mortar
2 iron tablets, for example, Boots' 'ferrous sulphate', 200 mg tablets
1M sulphuric acid, 100 cm³
0.005M potassium manganate(VII) (standardized)

Procedure

Full details are given in the *Students' book*.

There are some problems associated with the analysis of tablets because more than one reducing agent may be present. The most likely interference is from glucose but as glucose reacts only very slowly with acidified potassium manganate(VII) it is unlikely to affect the analysis significantly.

Some results obtained by a class of sixth-form students (using only one tablet instead of the two recommended) are as follows; average titres refer to the volume of 0.005M potassium manganate(VII) solution used:

Mass of tablet /g	Average titre /cm^3	Mass of FeSO$_4$ /mg	% Fe
0.595	4.96	188	11.8
0.642	5.00	190	10.9
0.580	4.95	187	11.8
0.597	4.65	177	10.9
0.493	4.60	174	13.0

The answers to the questions and exercises at the end of the experiment are as follows.

1 Excess 1M sulphuric acid is added for two reasons:

a to suppress the hydrolysis of the iron(II) ions which would otherwise take place (see experiment 16.2a);

b to supply the H^+(aq) ions needed for the reaction to proceed according to the equation.

2 The results appear to suggest that the 200 mg mentioned on the bottle label refers to the mass of *anhydrous* FeSO$_4$ in each tablet.

Iron tablets are taken medicinally to help combat anaemia, a condition in which the blood contains too little haemoglobin, a protein which contains iron in a complexed form (see the Background reading 'Micronutrients' at the end of this Topic in the *Students' book*).

Supporting homework

Answering questions at the end of this Topic in the *Students' book*. Recording the results of the experiments and working out the calculations.

Summary

At the end of this section, students should:

1 know some of the reactions of iron(II) and iron(III) compounds and of vanadium(II), (III), (IV), and (V) compounds;

2 have had more practice in predicting the outcome of reactions using standard electrode potentials;
3 know how to calculate the percentage of iron(II) ions using potassium manganate(VII) solution.

16.3
COMPLEX ION FORMATION
Objectives

1 To introduce the terms coordination number, stability constant, and mono-, bi-, hexa-, and poly-dentate ligand.
2 To investigate experimentally various complexes and compare the results with given stability constants.
3 To use the method of continuous variation to find the stoicheiometry of some complexes.
4 To carry out some complex ion preparations.

Timing

Six periods.

Suggested treatment

Overhead projection transparency number 118 will be useful. Students should read the introduction in the *Students' book* in advance before starting Experiment 16.3a, so that laboratory time is not wasted.

It is most important that students think about the questions at the end of each section of practical work and answer them before going on to the next section. Without this, the practical work will have little or no meaning.

EXPERIMENT 16.3a
An investigation of some copper(II) complexes

Each student, or pair of students, will need:
Rack of test-tubes
Dropping pipette
0.5M copper(II) sulphate, $10\,cm^3$
Concentrated hydrochloric acid, $5\,cm^3$
Concentrated ammonia solution, $15\,cm^3$
0.2M edta, $10\,cm^3$
0.1M sodium 2-hydroxybenzoate (sodium salicylate), $20\,cm^3$
0.1M 1,2-dihydroxybenzene (catechol) in 0.5M sodium hydroxide, $10\,cm^3$. (Note: care should be taken when preparing this solution as the solid is an irritant.)

Procedure

Full details are given in the *Students' book*.
Answers to questions in the *Students' book* are as follows.
1 The complex ion present is $Cu(H_2O)_4^{2+}$(aq).
2 It is pale blue.
3 The solution turns green.
4 The ligands in the complex ion are now Cl^-.
5 The solution is pale blue.
6 The complex ion present is $Cu(H_2O)_4^{2+}$(aq).
7 The equilibrium has been reversed.
8 The solution is now dark blue.
9 NH_3.
10 $Cu(NH_3)_4^{2+} > CuCl_4^{2-} > Cu(H_2O)_4^{2+}$.
11 Yes, the more stable complexes have the higher stability constants.
12 The edta-Cu(II) complex is pale blue, slightly darker than the aquo-complex.
13 It would turn pale blue.
14 Yes.
15 Blue, dark blue, green, green.
16 Blue $\longrightarrow$ dark blue $\longrightarrow$ green $\longrightarrow$ pale blue $\longrightarrow$ green.

At the end of this experiment the *Student's book* gives the formulae of the various ligands, and classifies them as monodentate, bidentate, or hexadentate.

In the next experiment students use the method of 'continuous variation' to find the stoicheiometry of some complexes. This experiment may need some explanation for those who have not used continuous variation before. It is possible, particularly in **1** and **2**, to use a colorimeter to estimate the point at which the maximum colour due to the complex is reached, but the maximum can be judged fairly reliably by eye. Some notes on the use of the colorimeter are given in Appendix 2 at the end of this book.

EXPERIMENT 16.3b
An investigation of the stoicheiometry of some complexes

Each student, or pair of students, will need:

2 burettes	2 racks of similar test-tubes

Access to solutions of:

0.1M nickel sulphate	0.1M copper(II) sulphate
0.005M potassium thiocyanate, KSCN	0.005M iron(III) chloride in hydrochloric acid
0.1M edta	0.1M phenylamine (aniline)

Procedure

Since the practical work is somewhat repetitive, the students could investigate one complex each and the results could be displayed for everyone to see.

Full details of the procedure are given in the *Students' book*.

In some cases it may help the students to decide on the correct empirical formula if a second experiment is devised in which $1\,cm^3$ portions of solution **A** are treated with 0.5, 1.0, $1.5\,cm^3$, etc., portions of solution **B**, and a judgment is again made of the mixture which gives the maximum colour of the complex.

The formulae of the complexes should work out to be:

$Ni(edta)$ $Fe(SCN)^{2+}$ $Cu(C_6H_5NH_2)_2^{2+}$

Work on the green complex of $Cu^{2+}(aq)$ with phenylamine has been reported in the following articles:

GAGGINI, P. AND MACPHERSON, J. 'An investigation into the formula of the copper(II)-phenylamine (-aniline) complex'. *School Science Review*. **191**, *55*, 332 (1973).

MEEK, E. G. 'Phenylamine-copper(II) sulphate(VI) complex: a possible project'. *School Science Review*. **204**, *58*, 489 (1977).

And also in WORSNOP, J. G. 'An analysis of a phenylamine complex'. *School Science Review*. **217**, *61*, 722 (1980).

The next experiment describes two possible preparations, but there is much scope for variation here. The following book, among many others, has useful collections of alternative experiments.

RENDLE, G. P., VOKINS, M. V. W., AND DAVIS, P. M. H. *Experimental Chemistry: a laboratory manual*. 2nd edition (1972). Edward Arnold.

EXPERIMENT 16.3c
The preparation of some compounds containing complexes

Each student, or pair of students, will need:

Safety glasses, one pair each
'Quickfit' (or similar) tap funnel
Multiple adaptor
'Quickfit' (or similar) flask, $50\,cm^3$
2 boiling tubes in rack
Bunsen burner
U-shaped tube and rubber stopper
(screw-cap adaptors may leak) to suit the adaptor and flask being used
Measuring cylinder, $10\,cm^3$
Beaker

Petri dish with lid
Access to rough balance
Concentrated hydrochloric acid, $14\,cm^3$
Copper turnings, 1.5 g
Granulated zinc, 2.5 g
Potassium dichromate(VI), 1 g
Sodium ethanoate, $10\,cm^3$
Thiourea (thiocarbamide), 4 g

Note. As mentioned in the *Students' book*, potassium dichromate(VI) is highly irritant to the skin, eyes, and respiratory system. Thiourea is very poisonous and must be handled with great care. Eye protection must be worn.

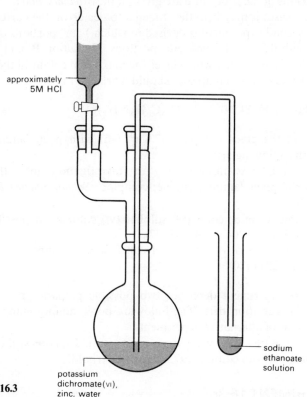

Figure 16.3

potassium dichromate(VI), zinc, water

approximately 5M HCl

sodium ethanoate solution

Procedure

Notes on method

1 Both experiments may require the products to be kept overnight, as both involve periods of waiting for a reduction or an oxidation to be complete. It is therefore highly desirable to have both preparations going on at the same time.

2 The product from the second experiment may separate as an oil rather than crystallizing. If the oily mixture is disturbed, crystallization should start.

3 Some tap funnels have long stems below the ground glass joints which prevent the funnels being fitted into the multiple adaptor as shown. In this case some rubber tubing can be sleeved over the stem and the stem then carefully fitted in place as shown in figure 16.4.

16.3 Complex ion formation 149

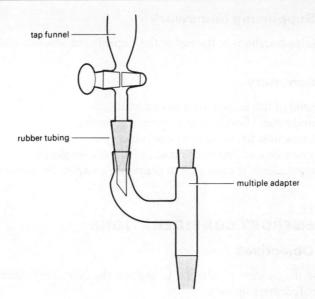

Figure 16.4

Full details of the procedure are given in the *Students' book*. The answers to the questions following the experiment are as follows.

Preparation 1
1 The relevant electrode potentials are:

$Zn^{2+}(aq)|Zn(s)$ $E^\ominus = -0.76 \text{ V}$
$Cr^{3+}(aq), Cr^{2+}(aq)|Pt$ $E^\ominus = -0.41 \text{ V}$
$[Cr_2O_7^{2-}(aq) + 14H^+(aq)], [2Cr^{3+}(aq) + 7H_2O(l)]|Pt$ $E^\ominus = 1.33 \text{ V}$

Application of the anti-clockwise rule reveals that zinc, Zn, should reduce $Cr_2O_7^{2-}(aq)$ to $Cr^{3+}(aq)$ and Cr^{3+} to $Cr^{2+}(aq)$.
2 The blue colour of chromium(II) changes immediately on exposure to air to the green colour of chromium(III).
3 It takes *much* longer for any change to occur in the red colour of chromium(II) ethanoate.

Preparation 2
1 The electronic structure of copper in the copper(I) ion is $[Ar]3d^{10}$, the $4s^1$ electron having been lost. This ion has a complete set of 3d orbitals, and colour is usually associated with incomplete d orbitals.
2 Thiourea is presumably a bidentate ligand so that each molecule occupies two positions in the octahedral complex.

Supporting homework

Answering questions at the end of the Topic in the *Students' book*.

Summary

At the end of this section students should:
1 understand how complex formation occurs;
2 know how to use stability constants;
3 know how to find the stoicheiometry of complexes;
4 have acquired some skill in preparing complex compounds.

16.4
ENTROPY CONSIDERATIONS

Objectives

To use the concept of entropy to explain the increasing stability of complexes with polydentate ligands.

Timing

One period.

Suggested approach

The *Students' book* covers the ground of this section. It could be studied at any time during section 16.3, or could follow it. The teacher could teach the material in class, if wished, but it is not intended that the treatment should be any deeper than that given in the *Students' book*.

16.5
TRANSITION ELEMENTS AS CATALYSTS

Objectives

1 To introduce the action of a catalyst in terms of potential energy.
2 To define catalysts as homogeneous or heterogeneous.
3 To explain the mechanism of heterogeneous catalysis in terms of adsorption and investigate this in more detail for ammonia synthesis.
4 To show how standard electrode potentials can be used to help choose a homogeneous catalyst.
5 To introduce Ellingham diagrams and their use in the iron and steel industry (Background reading).

Timing

One or two periods will be sufficient, together with some homework time.

Suggested treatment

No experiments are given for this section. The opportunity should be taken to revise ideas of catalysis introduced in section 14.4, and to provide some examples of the use of transition elements as catalysts.
 The Topic ends with two pieces of Background reading:
 1 Ellingham diagrams and the iron and steel industry.
 2 Micronutrients.
 Ellingham diagrams may well require some treatment in class. They form an excellent example of a practical application of free energy, and as such will repay the trouble taken to explain them.

Suggestions for homework

Answering questions at the end of the Topic in the *Students' book*.

Summary

At the end of this section, students should:
 1 know that transition elements are useful in both homogeneous and heterogeneous catalysis;
 2 know that the ability of transition elements to chemisorb gases underlies their usefulness as heterogeneous catalysts;
 3 know how to use standard electrode potentials to predict whether a particular transition compound might work as a catalyst.

After reading the Background reading at the end of the Topic, students should:
 1 be aware of the use of Ellingham diagrams;
 2 have learnt something of the extraction of iron and the production of steel;
 3 be aware of the importance of micronutrients.

ANSWERS TO PROBLEMS IN THE *STUDENTS' BOOK*

(A suggested mark allocation is given in brackets after each answer.)

  1a A and D. (2)
    bi According to the standard electrode potentials the manganate(VI) ion changes into the black manganese(IV) oxide and the purple manganate(VII) ion. (2)

ii Excess of hydroxide ion changes the electrode potential of equilibrium **B** to a less positive value, which makes it less likely that the reaction will occur. (2)
iii Mix potassium manganate(VIII) with manganese(IV) oxide in alkaline solution (concentrated). (3)
iv Disproportionation. (1)
c It should disproportionate into manganese(II) ions (colourless) and manganese(IV) oxide (black solid). (2)
d i Zn^{2+}, as there is no higher oxidation number of Zn.
Co^{2+}, as the electrode potential for oxidation to Co^{3+} is too highly positive. (2)
ii Chromium(II) ions in solution are very readily oxidized by the air to chromium(III) ions. (2)
iii Blue to pink. (2)

Total 18 marks

2a The relevant electrode potentials are:

$Zn^{2+}(aq)|Zn(s) \quad E^\ominus = -0.76\,V$
$Fe^{3+}(aq), Fe^{2+}(aq)|Pt \quad E^\ominus = +0.77\,V$

Application of the anticlockwise rule shows that Zn(s) should reduce $Fe^{3+}(aq)$ to $Fe^{2+}(aq)$. (3)
b The use of sulphuric acid minimizes hydrolysis of the iron(II) ions (see experiments 16.2a and 16.2c). (1)
c Using the method of experiment 16.2c, the percentage of iron by mass is 23.24%. (5)

Total 9 marks

3a $1s^2 2s^2 2p^6 3s^2 3p^6 3d^5$. (2)
b The shape should be octahedral (figure 16.5). (2)
c The oxidation number of iron in FeF_6^{4-} is +2. (1)
d The relevant electrode potentials are:

$I_2(aq), 2I^-(aq)|Pt \quad E^\ominus = +0.54\,V$
$Fe^{3+}(aq), Fe^{2+}(aq)|Pt \quad E^\ominus = +0.77\,V$

Application of the anticlockwise rule shows that $Fe^{3+}(aq)$ should oxidize $I^-(aq)$ to $I_2(aq)$ and be reduced to $Fe^{2+}(aq)$. (3)

Total 8 marks

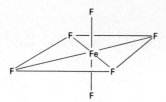

Figure 16.5

4 Mark by impression. **Total 10 marks**

5 The answer should centre on the fact that for the formation of a metal oxide the entropy change is positive, which means that ΔG becomes less and less negative as the temperature increases. For the formation of the oxides of carbon this is not so and, in fact, the formation of carbon monoxide from carbon and oxygen is the only common case of oxide formation which has a negative entropy change. The fact that carbon is a cheap, readily obtainable reducing agent is also important, of course. Mark by impression.

Total 10 marks

6a It would turn from pink to green. (2)
  b It would turn from green to pink since the hydrochloric acid reacts with the ammonia to give ammonium ions which are not ligands. (3)
  c There would probably not be much change of colour despite the likelihood of edta replacing the water molecules as ligands, because both complexes are pink. (3)
  d The entropy change for the reaction in which edta replaces water molecules is positive because six water molecules are being replaced by one molecule of edta. (2)

Total 10 marks

TOPIC 17
Synthesis: drugs, dyes, and polymers

OBJECTIVES

1 To develop knowledge of addition and condensation polymerization, the diazotization reaction, and other reactions useful in the synthesis of organic compounds.
2 To develop knowledge of some important synthetic polymers, their methods of manufacture, and their properties.
3 To introduce the organic chemistry of some dyes and some drugs.
4 To use analytical data, mass spectra, and infra-red spectra to identify unknown organic compounds.
5 To revise the main reactions of the functional groups and to develop knowledge of simple synthetic pathways.
6 To provide information about the industrial, historical, and social importance of appropriate selected compounds.

CONTENT

17.1 High polymers. The development of synthetic polymers; addition and condensation polymerization; preparation of some polymers and examination of their properties; choosing a polymer for use as a fibre; thermosetting polymers, elastomers, thermoplastics; commercial manufacture of some synthetic polymers; systematic, trivial, and trade names of some common polymers. Background reading: 1 'A problem for the polymer chemist'.
17.2 Dyes. Azo-dyes: experimental investigation of diazotization and coupling reactions in production of azo-dyes; investigation of the dyeing of different fabrics. Background reading: 2 'Dyestuffs: the origins of the modern organic chemical industry'.
17.3 Drugs. Preparations using 2-hydroxybenzoic acid to produce aspirin and oil of wintergreen. Background reading: 3 'Aspirin'; 4 'Drugs and medicines'.
17.4 The identification of organic compounds. Combustion analysis of organic compounds; the formation of derivatives to identify members of a homologous series; experiments to identify a carbonyl compound, and to identify three other unknown organic compounds.
17.5 Simple synthetic routes. Survey of useful reactions met earlier; a problem in synthesis (optional).

TIMING

This Topic should take four or five weeks (Experiment 17.5 is optional).

INTRODUCTION

This final Topic on organic chemistry is an opportunity to bring together all the important reactions studied in the previous Topics 9, 11, and 13, and to show how they can be connected together in simple synthetic pathways. There is, therefore, an opportunity to revise the reactions of the functional groups. Teachers should see that their students are quite clear about the differences and similarities between the structural formulae of the various functional groups, an area where errors of learning are common.

The new material in this Topic is mainly concerned with examples of important applications of organic chemistry, mostly included as Background reading; alternative examples could be presented to students if the teacher so wished. The problem in synthesis at the end of the Topic is an optional study; if students have had adequate experience of laboratory preparations, then a synthetic problem need not be attempted.

Students will be expected to know how to calculate empirical and molecular formulae from appropriate data, and to undertake the simple interpretation of spectra such as mass and infra-red spectra.

17.1
HIGH POLYMERS

Objectives

1 To introduce addition and condensation polymerization.
2 To prepare and investigate the properties of some synthetic polymers.
3 To outline the properties, uses, manufacture, and naming of some common synthetic polymers, and their advantages for the production of a wide range of goods.

Timing

About a week.

Suggested treatment

Overhead projection transparencies numbers 119 and 120 will be useful for this section.

Students should be reminded of the polymers they have already studied, poly(ethene) in Topic 9, polysaccharides in Topic 11, and nylon in Topic 13, before they are introduced to the methods used to manufacture the main synthetic polymers. The emphasis should be on the reactions involved rather than the industrial technology.

It is adequate to restrict the experimental work to a double period and only allow the students a limited choice of experiments. Detailed knowledge of each preparation is not expected.

EXPERIMENT 17.1a
The preparation of some synthetic polymers

Each student or pair of students will need:
1 *Poly(methyl 2-methylpropenoate) (Perspex)*
 Safety glasses
 Measuring cylinder, $25 \, cm^3$
 Test-tube, $150 \times 25 \, mm$
 Wooden splint
 Water bath ($400 \, cm^3$ beaker)
 Tripod, gauze, and Bunsen burner
 Di(dodecanoyl) peroxide (lauroyl peroxide), 0.1 g
 Methyl 2-methylpropenoate (methyl methacrylate), $5 \, cm^3$
2 *Poly(propenamide)*
 Safety glasses
 Protective gloves
 Beaker, $250 \, cm^3$
 Throw-away container, $250 \, cm^3$
 Heat-resistant mat
 Thermometer, 0–100 °C
 Potassium peroxodisulphate(VI) (potassium persulphate), 0.1 g
 Propenamide (acrylamide), 10 g
3 *Polyester resin*
 Access to fume cupboard
 Protective gloves
 Safety glasses
 Dropping pipette
 Oil bath
 Test-tube, $150 \times 25 \, mm$
 Thermometer, 0–250 °C
 Benzene-1,2-dicarboxylic anhydride (phthalic anhydride), 3 g
 Propane-1,2,3-triol (glycerol), $2 \, cm^3$
4 *Phenolic resin (a form of Bakelite)*
 Access to fume cupboard
 Test-tube, $150 \times 25 \, mm$
 40% aqueous methanal ('Formalin'), $1 \, cm^3$
 Sodium hydroxide, 0.2 g
 Phenol, 3 g

Procedure

Full details of the procedure are given in the *Students' book*. Students cannot be expected to perform more than one or two of the preparations in the time available, but the results can be considered in a general class discussion. Alternatively, some of the experiments could be done as a teacher's demonstration.

Students should be warned about the pungent odour and caustic nature of many of the monomers. Laboratories without efficient ventilation are advised to restrict the choice of experiments. Tin cans should be available for use as throwaway containers as laboratory glassware may prove impossible to clean.

EXPERIMENT 17.1b
An examination of the physical properties of some polymers

Each student or pair of students will need:
Safety glasses
Heavy metal plate, or tin lid
Liquid detergent
Metal spatula
Sharp knife
Test-tubes and rack
Poly(ethene) film
Solid granules of: Bakelite, nylon, Perspex, poly(chloroethene) (PVC), poly(ethene), and rubber

Procedure

Full details of the procedure are given in the *Students' book*.

Choosing the right fibre for a particular application

The discussion of the choice of a fibre for particular applications usually arouses considerable interest and results in an enhanced awareness of the diversity in properties of polymers and the skill needed to match those properties to the intended uses. The identification of the fibres is:

A wool   B cotton   C silk   D viscose rayon
E cellulose triacetate   F nylon 6   G polyester   H acrylic

The *Students' book* follows on with sections on the nature of polymers, the commercial manufacture of some synthetic polymers, and the names of polymers. The Background reading provides one case study of the production of a polymer for a particular application, but other case studies could be substituted.

158 Topic 17 Synthesis: drugs, dyes, and polymers

Supporting homework

Reading and making notes on the Background reading.

Summary

At the end of this section students should:
1 know what is meant by addition and condensation reactions;
2 know the laboratory and/or commercial preparations of various polymers;
3 know the main categories of polymers: thermoplastic, thermosetting, elastomer;
4 have some idea of how polymers are 'tailored' to fit particular applications.

17.2
DYES

Objectives

1 To introduce the diazotization reaction and its place in the production of dyes.
2 To show how the properties of the fabric affect the dye.
3 To introduce the place of dyes in the development of the organic chemical industry.

Timing

Four periods.

Suggested treatment

Overhead projection transparency number 121 will be useful in this section.
   It should be sufficient to carry out the experimental work and to ensure that students have studied the interpretations of reactions which are included as part of the *Students' book*.

EXPERIMENT 17.2a
The diazotization and coupling reactions

Each student or pair of students will need:

Safety glasses
3 beakers, $100 cm^3$
3 beakers, $250 cm^3$
Test-tubes and rack
Thermometer, 0–100 °C
Butylamine, $0.5 cm^3$

2M hydrochloric acid, $30 cm^3$
Naphthalen-2-ol (2-naphthol), 3 g
2M sodium hydroxide, $20 cm^3$
Sodium nitrite, 1.5 g
Phenylamine, $0.5 cm^3$
Ice

Procedure

Full instructions are given in the *Students' book*.

In reactions with nitrous acid, ammonia produces nitrogen gas, alkylamines produce nitrogen and a mixture of alcohol, alkene, and nitroalkane, but arylamines produce relatively stable diazonium salts.

In the 'blank' diazotization reaction mixture, a pale yellow crystalline solid will separate on adding the naphthalen-2-ol solution, due to nitrosation of naphthalen-2-ol. The same product may be observed in the butylamine reaction mixture if an excess of sodium nitrite is present. Acidification quickly turns this into a black tar, but this should only be demonstrated in a 'throw-away' container.

The electrophile, NO^+, is considered to be the attacking group in conditions of high acidity and in a sequence of steps the ion, $R\!-\!\overset{+}{N}\!\equiv\!N$, is formed. Unless this ion is stabilized in some way a nitrogen molecule will be formed:

$$R\!-\!NH_2 \xrightarrow{NO^+} R\!-\!\overset{+}{N}\!\equiv\!N \longrightarrow R^+ + N_2$$

The new ion R^+ will then take part in a variety of further reactions. In the case of arenes stabilization of $R\!-\!\overset{+}{N}\!\equiv\!N$ occurs by interaction involving delocalization with the benzene ring. The diazonium salts that are formed are useful reactive intermediates.

EXPERIMENT 17.2b
To investigate the dyeing of different fabrics

Each student or pair of students will need:
Safety glasses
Dyebath (400 cm^3 beaker)
Tongs
Dyes mixture (Direct Red 23, Disperse Yellow 3, Acid Blue 40), 0.05 g
25 cm^2 fabrics, e.g. cotton, nylon, cellulose ethanoate (acetate), polyester

Procedure

Full details are given in the *Students' book*. The mixture of three dyes, Direct Red 23, Disperse Yellow 3, and Acid Blue 40, has been selected to dye mainly one fabric in preference to other fabrics. Cotton is dyed red by Direct Red 23, polyester and cellulose ethanoate are dyed yellow by Disperse Yellow 3, nylon and wool are dyed green by a mixture of Acid Blue 40 and Disperse Yellow 3. The experiment demonstrates how a knowledge of the structure of dyes and fabrics enables the chemist to dye successfully the wide variety of fabrics that are manufactured.

Supporting material

Overhead projection transparency number 121 showing the molecular formulae of some dyes.
Booklets on dyes and dyeing are available from:
Customer Advice Bureau, Dylon International Limited, Worsley Bridge Road, Lower Sydenham, London SE26 5MD. (A large self-addressed envelope is requested.)
Public Relations Department, Organics Division, Imperial Chemical Industries plc, PO Box No. 42, Hexagon House, Blackley, Manchester M9 3DA.

Suggestions for homework

Reading the Background reading 'Dyestuffs: the origins of the modern organic chemical industry'.

Summary

At the end of this section students should:
 1 know what is meant by diazotization and coupling reactions and the mechanism behind them;
 2 have some knowledge of the dyes industry.

17.3 DRUGS

Objectives

1 To introduce the ethanoylation reaction.
2 To make students aware of the contribution of organic chemists to the development of modern drugs.

Timing

Four periods.

Suggested treatment

For this section overhead projection transparencies numbers 122, 123, and 124 will be useful.
Students are not expected as a first requirement to learn the details of the chemistry of aspirin, so it is suggested that the emphasis should be on the ethanoylation reaction. The opportunity can be taken to revise the chemistry of

carboxylic acid derivatives and the general methods of preparing esters, including the equilibrium nature of the reaction.

EXPERIMENT 17.3
Preparations using 2-hydroxybenzoic acid

Each student or pair of students will need:
Safety glasses
Apparatus for suction filtration
Beaker, 100 cm^3
2 conical flasks, 100 cm^3
Distillation apparatus (50 cm^3 pear-shaped flask, still head, and Liebig condenser)
Dropping tube
Ice bath
Measuring cylinder, 5 cm^3
Reflux apparatus consisting of Liebig condenser and 50 cm^3 pear-shaped flask
Separating funnel
Stirring rod
Thermometer, 0–250 °C
Ethanoic anhydride, 4 cm^3
2-hydroxybenzoic acid (salicylic acid), 11 g
Methanol, 15 cm^3
Phosphoric(v) acid, 85%, a few drops
0.5M sodium carbonate, 30 cm^3
Sodium sulphate, anhydrous
Sulphuric acid, concentrated, 2 cm^3
1,1,1-trichloroethane, 15 cm^3

Procedure

Full instructions are given in the *Students' book*.

The choice of phosphoric acid as catalyst for the manufacture of aspirin is based on the harmless nature of phosphoric acid and its low cost. It is instructive to compare the cost of different brands of aspirin tablets based on their aspirin content.

The Background reading on aspirin and on drugs and medicines could be set as homework and used as the basis for a discussion on the use and abuse of drugs in society today. As an alternative the exercise in decision making, 'To market a drug', could be used. It has been produced by the Royal Society of Chemistry Education Division (Scotland), and details are given below.

Supporting material

Booklets and newspaper cuttings on drugs and drug abuse.
'To market a drug', Scottish Council for Educational Technology, Dowanhill, 74 Victoria Crescent Road, Glasgow G12 9JN.

Suggestions for homework

Reading the Background reading 3 'Aspirin' and 4 'Drugs and medicines'.

Summary

At the end of this section students should:
1 know what the ethanoylation reaction is and how it is used in the preparation of aspirin and oil of wintergreen;
2 have some knowledge of the variety of drugs available, and their uses and abuses.

17.4
THE IDENTIFICATION OF ORGANIC COMPOUNDS

Objectives

1 To introduce the combustion analysis of organic compounds.
2 To introduce the use of derivatives (in this case with 2,4-dinitrophenylhydrazine) in the identification of individual members of a homologous series.
3 To practice identifying organic compounds using mass spectra, infra-red spectra, experimental evidence, and evidence from combustion analysis, and thereby revising earlier work.

Timing

About a week.

Suggested treatment

This section could be used as an opportunity to revise Topic 7 on the determination of molecular structures and Topic 4 on the mass spectrometer. An introduction to the calculation of empirical and molecular formulae should take a double period and the experimental work should take two double periods.

The *Students' book* starts with a description of the combustion analysis of an organic compound to determine the empirical formula, and the conversion of this to a molecular formula. Two sample problems are given, one is worked out and the answer is given in the other case.

The determination of the molecular formula of an organic compound is rarely sufficient to identify the compound unambiguously. The method of making a derivative and then identifying the compound from tables of data is described, and the reasons why 2,4-dinitrophenylhydrazine is a suitable reagent are given. In the next experiment students identify a carbonyl compound using this method.

EXPERIMENT 17.4a
The identification of a carbonyl compound

Each student or pair of students will need:
Safety glasses
Access to apparatus for suction filtration
Access to melting point apparatus with thermometer, 0–250 °C
Beaker
Melting point tubes
Test-tubes, 150 × 16 mm, and rack
Thermometer, 0–250 °C
Antibumping granules
Access to a selection of aldehydes and ketones (a table of possible compounds is given in the *Students' book*)
Brady's reagent, 2,4-dinitrophenylhydrazine; for preparation see instructions below
Dibutyl benzene-1,2-dicarboxylate (dibutyl phthalate)
Fehling's solutions A and B
Methanol
2M sulphuric acid

Procedure

Brady's reagent: 2,4-dinitrophenylhydrazine is conveniently kept as a stock solution in the form known as Brady's reagent. To prepare nearly 250 cm^3 dissolve 10 g of 2,4-dinitrophenylhydrazine in 20 cm^3 of concentrated sulphuric acid and add carefully, *with cooling*, 150 cm^3 of methanol. Warm gently to dissolve any solid and then add 50 cm^3 of water. Crystals usually form on the bottom of the reagent bottle during storage.

The preparation of Fehling's solutions A and B was described in Topic 11, on page 229 of *Teachers' guide I*.

The details of the procedure are given in the *Students' book*; table 17.5 gives the necessary physical data.

EXPERIMENT 17.4b
An investigation of three unknown organic compounds

Each student or pair of students will need:

Safety glasses
Access to melting point apparatus
Access to boiling point apparatus
Combustion spoon
Test-tubes and rack
Access to Unknown A (benzoic acid)
        Unknown B (propan-2-ol)
        Unknown C (butanone)

Brady's reagent (see Experiment 17.4a)
0.1M potassium dichromate(VI) solution
2M sodium carbonate solution
2M sulphuric acid
Calculator
Book of data

Procedure

Full details of the procedure are given in the *Students' book*.

The three investigations have been kept quite simple as they are intended only as elementary exercises and not as an introduction to a systematic scheme for the identification of organic functional groups. The investigations can be shortened by providing students with the results to some parts of the investigations, for example, the molecular formulae or the melting/boiling points.

Overhead projection transparencies numbers 125, 126, and 127, give the infra-red and mass spectra of unknowns A, B, and C.

Unknown A
1 Empirical formula $C_7H_6O_2$
2 Mass spectrum – see table 17.1
4d Melting point 121 °C
5 Benzoic acid

Unknown B
1 Empirical formula C_3H_8O
2 Mass spectrum – see table 17.2
4d Boiling point 83 °C
5 Propan-2-ol

Unknown C
1 Empirical formula C_4H_8O
2 Mass spectrum – see table 17.3
4c Melting point 115 °C
4d Boiling point 80 °C
5 Butanone

m/e	Groups commonly associated with the mass	Possible inference
39	$C_3H_3^+$	
50	$C_4H_2^+$	Aromatic compound
51	$C_4H_3^+$	C_6H_5-
77	$C_6H_5^+$	C_6H_5-
78	$C_6H_6^+$	C_6H_5-
105	$C_6H_5CO^+$	C_6H_5CO-
105	$C_8H_9^+$	
122	$C_6H_5CO_2H^+$	Molecular ion peak

Table 17.1

m/e	Groups commonly associated with the mass	Possible inference
27	$C_2H_3^+$	
29	$CHO^+, C_2H_5^+$	
39	$C_3H_3^+$	
41	$C_3H_5^+$	
43	CH_3CO^+	CH_3CO-
43	$C_3H_7^+$	C_3H_7-
45	$CH_2=\overset{+}{O}CH_3$ $CH_3CH=\overset{+}{O}H$	Some ethers and alcohols
60	$C_3H_7OH^+$	Molecular ion peak

Table 17.2

m/e	Groups commonly associated with the mass	Possible inference
26	$C_2H_2^+$	
27	$C_2H_3^+$	
28	$CO^+, C_2H_4^+, N_2^+$	
29	$CHO^+, C_2H_5^+$	
42	$C_2H_2O^+, C_3H_6^+$	
43	CH_3CO^+	CH_3CO-
43	$C_3H_7^+$	C_3H_7-
57	$C_4H_9^+$	C_4H_9-
57	$C_2H_5CO^+$	Ethyl ketone / Propanoate ester
72	$C_4H_8O^+$	Molecular ion peak

Table 17.3

Supporting homework

Answering questions from the end of the Topic in the *Students' book*.

Summary

At the end of this section students should:

1 be able to calculate empirical and molecular formulae from appropriate data;

2 be able to use simple data tables to interpret spectra such as infra-red and mass spectra;

3 be able to use experimental evidence to identify simple organic compounds.

17.5
SIMPLE SYNTHETIC ROUTES

Objectives

1 To provide opportunities to revise the reactions of the functional groups, and of free radical, electrophilic, and nucleophilic reagents.
2 To gain knowledge of simple synthetic pathways.
3 To gain further manipulative skill in organic chemistry.

Timing

About one and a half weeks (one week is optional).

Suggested treatment

The material presented in the *Students' book* has been kept simple so that the pattern of synthetic pathways can be identified without being obscured by too much detail. Students should, however, go back over the earlier Topics on organic chemistry and add any details of reagents and reaction conditions that they need to learn in conjunction with the reactions.

Experiment 17.5 is optional; it is intended as an alternative to the various laboratory preparations which were available in the course. Nevertheless if time is available it would be a valid activity for all students.

The *Students' book* suggests that practical books should be consulted for the students to plan a synthesis scheme of their own choice. It is felt that this is a valuable activity but teachers may have strong reservations about students actually executing their own schemes.

Due to inexperience, many schemes proposed by students may not be feasible and the successful execution of a scheme is as important as the planning. For this reason details of three schemes are given below which might be offered to students who in the teacher's estimation plan unrealistic or unrewarding schemes. The three schemes offered have a bias towards processes of industrial and social interest. It is hoped, for example, that the production of a dyestuff culminating in its application to some cloth will add more to the student's interest in his or her work than producing an obscure compound in a specimen tube.

Detailed descriptions of practical techniques (suction filtration, drying liquids, setting up apparatus) are not given, as it is felt that these are much better conveyed by a teacher's demonstration which can be related to the particular laboratory facilities available. Similarly, the apparatus list only contains the chemicals that are needed.

Teachers will appreciate that these schemes are examples only, and it is not intended that students should learn the details of these syntheses. Many alternative syntheses are available. Six were published in the first edition of *Teachers' guide II*, and others have been published in the *School Science Review*. The three syntheses published as part of this revised course are not considered easy to carry out but they are not readily accessible, except perhaps the indigo synthesis, and they have connections with other parts of the course.

Teachers who wish to follow up the social and economic stories connected with the syntheses described in the *Students' book* should find the following sources helpful.

For the steroid synthesis:
LEHMANN, P. A., BOLIVAR, A., and QUINTERO, R. 'Russell E. Marker, pioneer of the Mexican steroid industry'. *Journal of Chemical Education*, Vol. 50, No. 3, March 1973, pages 195–199.
KOLB, D. 'A pill for birth control'. *Journal of Chemical Education*, Vol. 55, No. 9, September 1978, pages 591–596.
For the benzocaine synthesis:
ANDREWS, G. and SOLOMON, D. (Editors). *The coca leaf and cocaine papers.* Harcourt Brace Jovanovich, 1975.

EXPERIMENT 17.5
A problem in synthesis

Each student or pair of students will need:
Reaction Scheme One: Preparation of indigo
2-aminobenzoic acid (anthranilic acid), 10 g
Concentrated hydrochloric acid
Cotton cloth, 10 cm^2
Monochloroethanoic acid, 7 g
Sodium carbonate, anhydrous, 16 g
Sodium disulphate(III), Na$_2$S$_2$O$_4$, 1 g
Sodium hydroxide, 12 g

Reaction Scheme Two: Preparation of Benzocaine
Calcium chloride, 2 g
Concentrated hydrochloric acid, 40 cm^3
Concentrated sulphuric acid, 60 cm^3
Ethanol, 170 cm^3
Ethoxyethane, 120 cm^3
Ice
4-nitromethylbenzene, 13.6 g
Pentane, 20 cm^3
Sodium chloride, 17 g
Sodium dichromate(VI) dihydrate, 40 g
1M sodium hydroxide, 240 cm^3
Sodium sulphate, anhydrous

Reaction Scheme Three: preparation of a steroid
95% aqueous ethanol, 8 cm^3
3.5% bromine reagent (see below), 12 cm^3
Cholesterol, 3 g
Ethanedioic acid, anhydrous (see below), 0.1 g
Ethanoic acid, pure, 60 cm^3
Ethoxyethane, 40 cm^3
Ice
Methanol, 50 cm^3
Pyridine, 1.4 cm^3
4% sodium dichromate(VI) reagent (see below), 40 cm^3
Sodium hydrogencarbonate, 0.5 g
Sodium sulphate, anhydrous
Zinc dust, 0.8 g

1 Bromine reagent. Enough bromine reagent for five experiments is obtained by adding 0.5 g anhydrous sodium ethanoate (see note 3) and 2.05 cm^3 bromine to 60 cm^3 of pure ethanoic acid.

2 Sodium dichromate(VI) reagent. Enough for five experiments is obtained by adding 8 g of sodium dichromate(VI) dihydrate to 200 cm^3 of pure ethanoic acid.

3 Hydrated sodium ethanoate and hydrated ethanedioic acid can be dehydrated by gentle heating in crucibles.

Procedure

Full details of the procedure for each reaction scheme are given in the *Students' book*.

Supporting homework

Answering problems from the end of the Topic in the *Students' book*.

Summary

At the end of this section students should:
1 have revised and brought together their knowledge of reactions from different Topics;
2 have gained further experience in organic chemistry techniques (optional).

ANSWERS TO PROBLEMS IN THE *STUDENTS' BOOK*

(A suggested mark allocation is given in brackets after each answer.)

1 i Homolytically and ii heterolytically.   (1 mark each)
   i Free radicals, ii nucleophiles, and iii electrophiles.
   i Substitution, ii addition, and iii elimination.
                               Subtotal 8 marks
   Suitable examples.   (8)
                               Total 16 marks

2a Free radical substitution.   (2)
 b Electrophilic addition.   (2)
 c Electrophilic substitution.   (2)
 d Oxidation.   (1)
 e Nucleophilic addition.   (2)
 f Elimination.   (1)
 g Nucleophilic substitution (in fact addition-elimination).   (2)
                               Total 12 marks

3 The reactions of benzene as primarily electrophilic substitutions (4 are important) should be contrasted with electrophilic addition reactions of cyclohexene (several should be considered). The two addition reactions of benzene with H_2 and Cl_2 which do not involve electrophiles should be mentioned.
   Similarities in reactions arise from the behaviour of amines as bases, ligands, and nucleophiles because of the lone-pair of electrons on nitrogen. Differences are due to the lower availability of the lone-pair in phenylamine on account of delocalization with the ring.
                               Total 20 marks

4a React with HBr (from KBr + concentrated H_2SO_4). Nucleophilic substitution.   (2)
 b React with hot, concentrated alcoholic solution of KOH. Elimination. (2)
 c Reflux with an alcoholic solution of NaCN. Nucleophilic substitution. (2)
 d React with CH_3OH and concentrated H_2SO_4 catalyst. Esterification. (Nucleophilic substitution, in fact, addition-elimination reaction).   (2)
 e React with $LiAlH_4$. Reduction.   (2)
 f React with PCl_3 or PCl_5. Nucleophilic substitution, in fact, addition-elimination reaction.   (2)
                               Total 12 marks

5a 2-methylbutan-1-ol. (1)

bi $CH_3-CH_2-\underset{\underset{CH_3}{|}}{CH}-CH_2I$ (1)

ii Red phosphorus and I_2. (KI and H_2SO_4 is not an acceptable answer.) (1)

iii $CH_3-CH_2-\underset{\underset{CH_3}{|}}{CH}-CH_2NH_2$ or $CH_3-CH_2-\underset{\underset{CH_3}{|}}{CH}-CH_2NH_3^+I^-$ (1)

iv $(CH_3-CH_2-\underset{\underset{CH_3}{|}}{CH}-CH_2)_2NH$ or $(CH_3-CH_2-\underset{\underset{CH_3}{|}}{CH}-CH_2)_3N$

or $(CH_3-CH_2-\underset{\underset{CH_3}{|}}{CH}-CH_2)_4N^+I^-$ (2)

ci Oxidize with $K_2Cr_2O_7/H_2SO_4$. (2)

ii $LiAlH_4$ (1)

di (2)

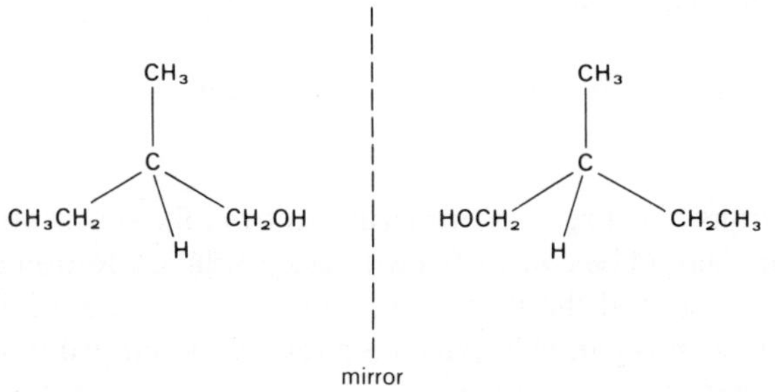

[Mirror image structures of 2-methylbutan-1-ol showing chirality at carbon attached to CH₂OH, with CH₃, CH₃CH₂, H, CH₂OH on one side and mirror image on the other]

ii The carbon atom attached to the —CH_2OH group is chiral. (1)

iii By using a polarimeter. One form rotates the plane of plane polarized light clockwise, the other anti-clockwise by the same amount. (2)

Total 14 marks

6a **B** Pass vapour over hot Al_2O_3 or treat liquid with phosphoric(v) acid.
  C $K_2Cr_2O_7/H_2SO_4$.
  F Na.
  I CH_3CO_2H and concentrated H_2SO_4 catalyst; CH_3COCl or $(CH_3CO)_2O$. (4)

b **A** Esterification. (Nucleophilic substitution, in fact, addition-elimination.)
  B Elimination.
  C Oxidation.
  I Esterification. (Nucleophilic substitution, in fact, addition-elimination.) (4)

c Formation of **F**. (1)

d Aqueous sodium hydroxide. (1)

e CH$_3$—CH—CO$_2^-$ NH$_4^+$
       |
      OH

Heat slowly to dehydrate. (2)

f G is formed by esterification between two molecules of 2-hydroxypropanoic acid. (1)

g Negative test with Fehling's, Tollen's, or Benedict's reagents. (1)

         Br   Br
         | |

hi CH$_3$—CHCO$_2$H or CH$_2$—CH$_2$CO$_2$H (2)

Markovnikov's rule suggests the first structure is more likely.

ii (2)

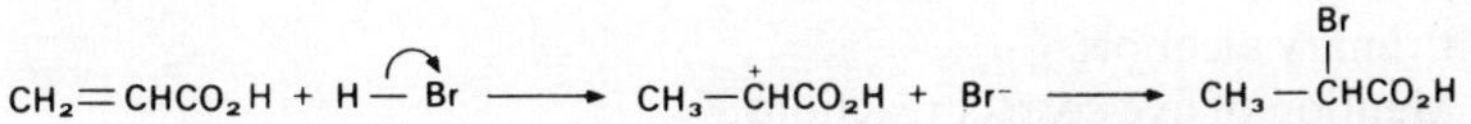

$$CH_2=CHCO_2H + H-Br \longrightarrow CH_3-\overset{+}{C}HCO_2H + Br^- \longrightarrow CH_3-\underset{|}{\overset{Br}{C}}HCO_2H$$

Total 18 marks

7a CH$_3$—CH$_2$—CH$_2$—CH$_2$Br   CH$_3$—CH$_2$—CH$_2$—C(=O)—OH (2)

b Yes. (1)

c Carboxylic acids. (1)

d Butan-1-ol. (1)

e Yes, by hydrolysing with NaOH solution. (2)

f CH$_3$—CH$_2$—CH$_2$—CH$_2$Br + OH$^-$ $\xrightarrow{\text{NaOH}}$

               CH$_3$—CH$_2$—CH$_2$—CH$_2$OH + Br$^-$ (2)

CH$_3$—CH$_2$—CH$_2$—CH$_2$OH + O$_2$ $\xrightarrow[\text{H}_2\text{SO}_4]{\text{K}_2\text{Cr}_2\text{O}_7}$

               CH$_3$—CH$_2$—CH$_2$—C(=O)—OH + H$_2$O (2)

Total 11 marks

8a $CH_3-CH_2-CH_2OH$ $CH_3-CH_2-CH_2-\overset{\overset{O}{\|}}{C}-OH$ (2)
b No. (1)
c Nucleophilic substitution to introduce $-CN$ group. (1)
d Yes. $CH_3-CH_2-CH_2-CN$ (2)
e $CH_3-CH_2-CH_2-OH \xrightarrow{HBr} CH_3-CH_2-CH_2Br \xrightarrow{KCN}$

$CH_3-CH_2-CH_2CN \xrightarrow{H_2O} CH_3-CH_2-CH_2-\overset{\overset{O}{\|}}{C}-OH$ (4)

Total 10 marks

9a Primary alcohols. (1)
b Mention of five correct reactions. (5)
c $CH_3-CH_2-\underset{\underset{Br}{|}}{CH}-CH_2Br$ (1)

d Addition reaction. (1)
e But-1-ene. $CH_3-CH_2-CH=CH_2$ (2)
f $CH_3-CH_2-CH_2-CH_2OH \xrightarrow[\text{or } Al_2O_3]{H_3PO_4}$

$CH_3-CH_2-CH=CH_2 + H_2O \xrightarrow{Br_2} CH_3-CH_2-\underset{\underset{Br}{|}}{CH}-CH_2Br$ (4)

Total 14 marks

10a $C_2H_5OH \xrightarrow[H_2SO_4]{K_2Cr_2O_7} CH_3-CO_2H$

$CH_3-CO_2H + C_2H_5OH \xrightarrow{H^+} CH_3-CO_2C_2H_5 + H_2O$ (4)

b $C_2H_5OH \xrightarrow[H_2SO_4]{K_2Cr_2O_7} CH_3-CO_2H$

$CH_3-CO_2H + CH_3-CH_2-CH_2-CH_2OH \xrightarrow{H^+}$
$\hspace{6cm} CH_3-CO_2C_4H_9 + H_2O$

$CH_3-CO_2H + CH_3-CH_2-\underset{\underset{}{}}{\overset{\overset{OH}{|}}{CH}}-CH_3 \xrightarrow{H^+}$
$\hspace{6cm} CH_3-CO_2C_4H_9 + H_2O$ (4)

c C₆H₅−C(=O)−OH $\xrightarrow{\text{PCl}_3 \text{ or PCl}_5}$ C₆H₅−C(=O)−Cl

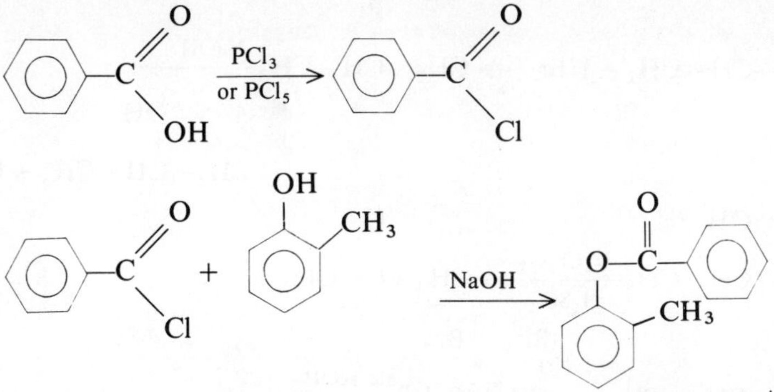

C₆H₅−C(=O)−Cl + 2-methylphenol $\xrightarrow{\text{NaOH}}$ 2-methylphenyl benzoate + HCl (4)

d $CH_3-CH_2-CH_2OH \xrightarrow{HBr} CH_3-CH_2-CH_2Br + H_2O$

$\xrightarrow{NH_3} CH_3-CH_2-CH_2NH_2 + HBr$ (4)

e $CH_3-CH_2-CH_2OH \xrightarrow{Al_2O_3} CH_3-CH=CH_2 + H_2O$

$CH_3-CH=CH_2 + HBr \longrightarrow CH_3-\underset{|}{\overset{Br}{C}}H-CH_3$

$CH_3-\underset{|}{\overset{Br}{C}}H-CH_3 + H_2O \xrightarrow{NaOH} CH_3-\underset{|}{\overset{OH}{C}}H-CH_3 + HBr$ (4)

Total 20 marks

11a $CH_3-CH_2Br \xrightarrow{KCN} CH_3-CH_2CN \xrightarrow{Ni/H_2} CH_3-CH_2-CH_2NH_2$ (3)

b $CH_3-CH_2-CH_2NH_2 + CH_3COCl \longrightarrow$
$\qquad\qquad CH_3-CH_2-CH_2NHCOCH_3 + HCl$ (2)

c C₆H₅−NH₂ $\xrightarrow{\text{HNO}_2 / \text{HCl}}$ C₆H₅−N⁺≡NCl⁻

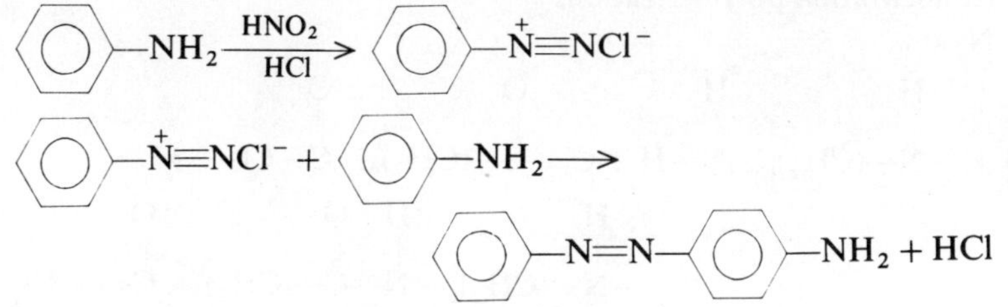

C₆H₅−N⁺≡NCl⁻ + C₆H₅−NH₂ ⟶

C₆H₅−N=N−C₆H₄−NH₂ + HCl   (3)

d $CH_3-CH=CH_2 + HBr \longrightarrow CH_3-\underset{\underset{Br}{|}}{CH}-CH_3 \xrightarrow{NaOH}$

$CH_3-\underset{\underset{OH}{|}}{CH}-CH_3 + HBr$

$CH_3-\underset{\underset{OH}{|}}{CH}-CH_3 \xrightarrow[H_2SO_4]{K_2Cr_2O_7} CH_3-\underset{\underset{O}{\|}}{C}-CH_3$ (4)

e $CH_2=CH_2 + Br_2 \longrightarrow \underset{\underset{Br}{|}}{CH_2}-\underset{\underset{Br}{|}}{CH_2} \xrightarrow{alc.\ KOH}$

$H-C\equiv C-H + 2HBr$ (3)

Total 15 marks

12a i Acidified $K_2Cr_2O_7$. Oxidation. (2)
  ii CH_3COCl. Esterification. (2)
 b Any two suitable reactions. (4)
 c Plastics. (1)
 d By recrystallizing from a suitable solvent (hot water). (1)
 e The direct route: only CO_2 is consumed. (2)
 f Disprin is fully ionized. (1)

Total 13 marks

13a i Alcoholic solution of KCN.
  ii H_2 and Ni catalyst.
  iii H_2 and Ni catalyst.
  iv PCl_3 or PCl_5. (4)
 b Condensation polymerization. (1)
 c Nylon. (1)
 d $H-\underset{\underset{H}{|}}{N}-(CH_2)_6-\underset{\underset{H}{|}}{N}-H + Cl-\overset{\overset{O}{\|}}{C}-(CH_2)_4-\overset{\overset{O}{\|}}{C}-Cl \longrightarrow$

  $-\underset{\underset{H}{|}}{N}-(CH_2)_6-\underset{\underset{H}{|}}{N}-\overset{\overset{O}{\|}}{C}-(CH_2)_4-\overset{\overset{O}{\|}}{C}- + HCl$ (2)

 e Warm to soften and then draw or extrude material; or 'cold drawing'. (2)
 f i The crystalline form has the higher elasticity.
  ii The crystalline form has a higher tensile strength.
  iii The crystalline form is isotropic and therefore rotates the plane of polarization of polarized light. (2)

Total 12 marks

Answers to problems in the Students' book   175

14 Essay question: mark by impression. **Total 20 marks**

15 Essay question: mark by impression. **Total 20 marks**

16a Low temperature (below 5 °C), excess acid, and $NaNO_2$. (2)
  bi The energetic instability is because of the tendency to form the very stable molecule N_2. (2)
  ii Diazonium salts are stabilized by delocalization with the benzene ring. (2)
  ci The phenate ion has been formed under alkaline conditions. (1)
  ii Electrophile. (1)
  iii Substitution of H by the electrophile. (1)
  d For the identification of aldehydes and ketones by preparation of crystalline derivatives. (1)
  ei Nitrogen is more electronegative than carbon and carries a full positive charge. The electrons of the azo group are delocalized with the benzene ring and electrons are withdrawn from it on account of the positive charge on the nitrogen atom. (2)
  ii The H atom is normally substituted. N_2^+ is expected when diazonium salts react because of the stability of N_2 as a leaving group. (2)
  iii There is nucleophilic attack by H_2O or OH^- at the C atom carrying the azo group. (2)
  iv Iodobenzene by nucleophilic substitution. (1)

Total 17 marks

17a 2 moles. (1)

  b (1)

  c By adding 2 moles of HCl (Markovnikov addition). (1)
  d By refluxing with aqueous NaOH. (1)
  ei Tertiary alcohol groups would not be oxidized without destruction of the molecule. (2)

  ii , but other dienes are possible. (2)

Total 8 marks

18a Infra-red spectroscopy and any suitable chemical reaction. (2)
 b A reducing agent such as $NaBH_4$. (2)
 c By vapour phase reduction with Ni/H_2. (2)
 d One —OH group has phenol reactions and the other has reactions of an alcohol. For a good list of reactions: (6)
 e Radioactive labelling of one of the carbon atoms in ethanoic acid, followed by biosynthesis and destructive analysis of the steroid. (2)
 Total 14 marks

19a III and V are geometric isomers. (2)
 b I ⟶ II, HBr.
  II ⟶ III, aqueous NaOH.
  III ⟶ IV, $K_2Cr_2O_7$ and H_2SO_4.
  IV ⟶ V, $NaBH_4$. (4)
 ci Atropine contains an ester linkage. Hydrolysis of the ester would form tropine. (2)
 ii(1) Hydrolyse both ester groupings.
  (2) Decarboxylate.
  (3) Oxidize alcohol to ketone. (4)
 Total 12 marks

TOPIC 18
The Periodic Table 5: the elements of Groups III, IV, V, and VI

OBJECTIVES

1 To study the patterns in behaviour of oxides and chlorides of some p-block elements.

2 To study the variable oxidation number of some of the elements of the p-block and to provide further opportunities to use standard electrode potentials predictively.

3 To provide experimental investigations of the behaviour in 1 and 2 above.

4 To study commercial applications of some elements and compounds from the p-block, especially the extraction of aluminium, the use of silicon and germanium as semiconductors, and the manufacture of sulphuric acid.

5 To study the main sources and effects of pollution caused by lead in the environment.

CONTENT

18.1 Some general features of the chemistry of the p-block elements. Electronic structure and the expansion of the octet; hydrolysis of p-block halides and hydrides; acid-base properties of p-block oxides.

18.2 Group III: boron and aluminium. Experimental investigation of boric acid and aluminium hydroxide; comparison of their acid-base properties; reactions of chlorides of boron and aluminium. Background reading: 1 'The extraction of aluminium'.

18.3 The elements of Group IV. Metallic character and oxidation numbers; comparison of oxides and chlorides of group using data tables; preparation and comparison of some oxides of tin and lead; estimation of percentage purity. Background reading: 2 'The chemist and micro-electronics': 3 'Lead pollution'.

18.4 Group V: the properties of some nitrogen compounds. Experimental investigations of some nitrogen compounds, related to standard electrode potentials.

18.5 Group VI: the properties of some sulphur compounds. Variable oxidation number; experimental investigations of sulphur compounds, related to standard electrode potentials; the manufacture of sulphuric acid.

TIMING

About two and a half weeks.

INTRODUCTION

Throughout the Topic, since it is the last topic in the course, opportunities should be taken to revise parts of earlier topics and in several places there are specific suggestions about revision which might be undertaken.

18.1
SOME GENERAL FEATURES OF THE CHEMISTRY OF THE p-BLOCK ELEMENTS

Objectives

1 To introduce the concept of 'expansion of the octet' and relate this to the properties of the p-block elements.
2 To study trends in the properties of the chlorides and oxides of the p-block elements.

Timing

One or two double periods, depending on the previous experience and ability of the students.

Suggested treatment

Overhead projection transparency number 128 will be useful.
The section in the *Students' book* deals with some general points about the elements of the p-block as a whole. The students should study these points at this stage so that they can look for examples of them amongst the descriptive chemistry in the rest of the Topic.

It should be borne in mind that the treatment in Groups III and IV emphasizes the acid-base properties of oxides and hydroxides and the hydrolysis of chlorides, whereas in the later groups the interest shifts more to redox reactions.

As part of the discussion of the chlorides of p-block elements students could revise work on the structures of the chlorides in Topic 5.

Supporting homework

Revision of Topic 5.

Summary

At the end of this section students should:
1 know what is meant by 'expansion of the octet';
2 be able to explain the mechanism of the hydrolysis of chlorides and hydrides;
3 know what is meant by basic, acidic, and amphoteric oxides;
4 know the pattern of acid-base behaviour of the p-block oxides.

18.2
GROUP III: BORON AND ALUMINIUM
Objectives

1 To look in more detail at the reaction of the hydroxides and chlorides of these two p-block elements.
2 To provide information about the industrial extraction of aluminium.

Timing

Two double periods will be required for the practical work, and some homework time.

Suggested approach

This section serves to illustrate the increase in basic properties of oxides and hydroxides on going down a group. Boric acid (which is the nearest boron gets to forming a hydroxide) is entirely acidic whereas aluminium hydroxide is amphoteric. The molecular structure of boric acid is given in the *Students' book*.

EXPERIMENT 18.2
Boric acid and aluminium hydroxide

Each student or pair of students will need:

Test-tubes, 100 × 16 mm, and test-tube rack
Corks to fit test-tubes
Boiling tube, 100 × 25 mm
2 beakers, 100 cm^3
Buchner filtration apparatus
Ice
'Pure water'
Sodium chloride
Aluminium sulphate, $Al_2(SO_4)_3 \cdot 16H_2O$, 0.5 g

Borax, 2.5 g
Concentrated hydrochloric acid
1M hydrochloric acid
0.1M hydrochloric acid
1M sodium hydroxide solution, 3 cm^3
0.1M sodium hydroxide solution
Full-range Indicator solution
Methyl red solution

Procedure

Full details are given in the *Students' book*. The following notes may be helpful.

Part 3 Comparison of the acid-base properties of boric acid and aluminium hydroxide

i The aluminium hydroxide should dissolve easily in both 1M hydrochloric acid and 1M sodium hydroxide. It is therefore an amphoteric hydroxide.

ii The contents of test-tubes A and B should turn red with the addition of only a drop or two of 0.1M hydrochloric acid. The contents of test-tube D require only a drop or two of 0.1M sodium hydroxide to produce a blue colour. Test-tube C however requires several drops. Boric acid is therefore not reacting with the hydrochloric acid but it is reacting with the sodium hydroxide.

Background reading

This section of the *Students' book* ends with a piece of Background reading on the extraction of aluminium.

Supporting homework

Completing the 'Personal work' suggested in the *Students' book*. Reading the Background reading 'The extraction of aluminium'.

Summary

At the end of this section students should:

1 know the reactions of the hydroxides, oxides, and chlorides of boron and aluminium they have met in this section, and know how they relate to the general trends in the behaviour of the p-block elements;

2 be aware of the method of extraction of aluminium.

18.3
THE ELEMENTS OF GROUP IV

Objectives

1 To outline the main trends in behaviour of the Group IV elements and their compounds.

2 To provide further practice in preparation of compounds.

3 To compare the properties of tin(IV) oxide and lead(IV) oxide and relate these to the generalizations about oxides made in section 18.1.

4 To provide further practice in estimating the percentage purity of a compound.

18.3 The elements of Group IV

5 To introduce students to the properties of elemental semiconductors, their extraction, and their manufacture into transistors.
6 To introduce students to the main causes and dangers of lead pollution.

Timing

Two double periods for the experimental work.

Suggested treatment

The introduction to this section will require either research in books on the part of the student or about one extra double period of teaching time. As pointed out, the intention of this section is to find examples of generalizations made about oxides and chlorides in section 18.1.

EXPERIMENT 18.3a
Preparation of some oxides of tin and lead

Each student or pair of students will need:
Safety glasses, 1 pair each
Boiling tube, 100 × 25 cm
Buchner filtration apparatus
Bunsen burner
Evaporating basin
400 cm^3 beaker
Access to fume cupboard
Concentrated nitric acid
Granulated tin, 0.5 g
Lead foil, 1 g
15% (2M) sodium chlorate(I) solution (sodium hypochlorite), 6 cm^3
2M sodium hydroxide solution

Procedure

Warn students that nitric acid is highly corrosive and that they should be particularly careful to avoid spillages. The tin(IV) oxide preparation should be carried out in a fume cupboard because of the evolution of nitrogen dioxide. Safety glasses should be worn throughout this experiment.
Full details of the procedure are given in the *Students' book*.

EXPERIMENT 18.3b
Comparison of tin(IV) oxide and lead(IV) oxide

Each student or pair of students will need:
Bunsen burner
Test-tubes, and test-tube rack
1M hydrochloric acid
0.1M potassium iodide solution
Lead(IV) oxide and tin(IV) oxide from Experiment 18.3a

Procedure

Full details are given in the *Students' book*.

Lead(IV) oxide oxidizes chloride ions to chlorine and lead(II) chloride crystals are deposited on cooling. With tin(IV) oxide there is no perceptible reaction. Lead(IV) oxide readily oxidizes acidified potassium iodide solution whereas tin(IV) oxide has no effect.

Answers to questions
1 Lead(IV) oxide is a much better oxidizing agent than tin(IV) oxide.
2 It was more difficult to oxidize lead to the +4 state than it was to oxidize tin.

EXPERIMENT 18.3c
An estimation of the percentage purity of lead(IV) oxide

Each student or pair of students will need:
Burette and funnel
Conical flask for titration
250 cm^3 conical flask (stoppered)
10 cm^3 pipette and filler
100 cm^3 volumetric flask
Weighing dish or watch glass
Lead(IV) oxide as prepared in Experiment 18.3a
2M hydrochloric acid, 50 cm^3
Potassium iodide, 1 g
1% starch solution (freshly made)
0.05M sodium thiosulphate solution

Procedure

Full details are given in the *Students' book*.

In trials a sample of lead(IV) oxide was found to be 85% pure. Much depends on how dry the sample of lead(IV) oxide is when it is weighed.

Background reading

This section of the *Students' book* ends with two pieces of Background reading. The first, 'The chemist and micro-electronics', describes the use of silicon dioxide in the manufacture of transistors and integrated circuits. The second, 'Lead pollution', discusses the widespread and dangerous effects of lead pollution, and describes in some detail the main sources of lead in the environment. The following information may help the teacher in a class discussion.

Lead is present in large quantities in the Earth's crust; from there some of it reaches the soil, the air and living organisms through natural processes. But this natural distribution is insignificant when compared with the distribution which results from man's activities.

The cumulative effect of centuries of use is that lead is now one of the most widely dispersed of environmental pollutants. It has been estimated that the worldwide annual emission of lead to the atmosphere is about 25 thousand tonnes from natural sources compared with 450 thousand tonnes from man's activity. See figure 18.1. Moreover, so far as is known, there is no innocuous form into which it can be converted in the environment. Once in the environment lead and its compounds do not move readily through natural pathways to remoter locations such as the ocean bed. Thus not only is there widespread human exposure to lead today but future generations too will be exposed to the lead which is already in the environment and is being added to all the time.

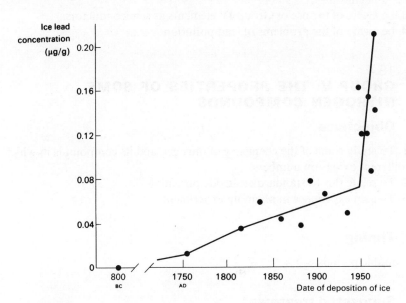

Figure 18.1
Lead content of annual ice layers in Greenland.

Besides lead there are other heavy metals in the environment which are potentially hazardous, notably mercury and cadmium. But what makes lead different from other heavy metals is the very much greater scale on which it is used, dispersed and accumulated. This factor, when coupled with its known and postulated effects on health in amounts which may be encountered in the environment and found in the human body, makes lead unique amongst environmental pollutants.

Supporting homework

Answering questions at the end of the topic in the *Students' book*.
Reading the Background reading 'The chemist and microelectronics' and 'Lead pollution'.

Summary

At the end of this section students should:
 1 know how metallic properties and oxidation numbers of the Group IV elements vary down the group;
 2 know the reactions of the various Group IV oxides and chlorides which they have met, and how they fit in with the overall pattern of behaviour of the oxides and chlorides;
 3 be aware of the use of Group IV elements as semiconductors;
 4 be aware of the problems of lead pollution.

18.4
GROUP V: THE PROPERTIES OF SOME NITROGEN COMPOUNDS

Objectives

 1 To study some of the chemistry of nitrogen and its compounds in which it has different oxidation numbers.
 2 To relate this to standard electrode potentials.
 3 To gain experience in planning experiments.

Timing

Two double periods.

Suggested treatment

Overhead projection transparency number 129 will be useful.

18.4 Group V: the properties of some nitrogen compounds

The main emphasis in this section is on the redox properties of some nitrogen compounds.

Outline instructions are given in the *Students' book* but opportunities may be taken for students to plan the details of the experiments for themselves, possibly for the purpose of practical assessment.

EXPERIMENT 18.4a
Some reactions of nitric acid, HNO_3, in concentrated solution

Each student or pair of students will need:
Safety glasses, 1 pair each
Test-tubes, 100 × 16 mm and test-tube rack
Concentrated nitric acid
Copper turnings
0.1M potassium iodide solution
Sulphur dioxide solution

Procedure

Safety glasses should be worn for this experiment.

Outline instructions only are given in the *Students' book*. The results of the experiments should be as follows.

1 The products are Cu^{2+}(aq) and N_2O_4/NO_2(g).

2 Sulphur dioxide is oxidized to sulphate ion. This may be confirmed by using the barium chloride test for a sulphate.

3 Iodine and a mixture of oxides of nitrogen are produced.

EXPERIMENT 18.4b
Some reactions of nitrous acid, HNO_2, in dilute solution

Each student or pair of students will need:
Safety glasses, 1 pair each
Bunsen burner
Test-tubes, 100 × 16 mm, and test-tube rack
1M hydrochloric acid
0.1M potassium iodide solution
0.1M potassium manganate(VII) solution
Sodium nitrite (CARE: very poisonous)

Procedure

Safety glasses should be worn for this experiment.

Outline instructions only are given in the *Students' book*.
The results should be as follows.

1 Iodine and oxides of nitrogen are produced when nitrous acid and potassium iodide solution are mixed.

2 Nitrous acid reduces manganate(VII) to manganese(II).

EXPERIMENT 18.4c
Some reactions of ammonia solution

Each student or pair of students will need:
Test-tubes, 100 × 16 mm, and test-tube rack
1M ammonia solution
1M ammonium chloride solution
0.1M copper(II) sulphate solution
0.1M lead(II) sulphate solution
0.1M zinc sulphate solution
0.1M magnesium sulphate solution

Procedure

Outline instructions are given in the *Students' book*.
The following notes may be useful.

1 Lead(II) hydroxide is precipitated when ammonia reacts with lead(II) ions.

2 With copper(II) ions the pale blue precipitate first produced dissolves to give the complex ion $Cu(NH_3)_4^{2+}$ (see Topic 16).

3 A similar reaction occurs with zinc ions to give the complex ion $Zn(NH_3)_4^{2+}$.

In the second part of the experiment with magnesium ions, the presence of ammonium ions, $NH_4^+(aq)$, prevents the precipitation of magnesium hydroxide by displacing the equilibrium

$$NH_3 + H_2O \rightleftharpoons NH_4^+ + OH^-$$

to the left and so reducing the concentration of $OH^-(aq)$.

Supporting homework

Answering question 7 at the end of the Topic in the *Students' book*.
Revising section 16.2 on variable oxidation numbers.

Summary

At the end of this section students should know some of the reactions of nitrogen and its compounds, and how these are related to oxidation numbers and standard electrode potentials.

18.5
GROUP VI: THE PROPERTIES OF SOME SULPHUR COMPOUNDS

Objectives

1 To study some of the chemistry of sulphur, using ideas already encountered in the course.
2 To inform students of the commercial importance of sulphuric acid and its methods of manufacture.

Timing

Two double periods.

Suggested approach

Overhead projection transparencies numbers 39 and 72 will be found useful.

Students can be introduced to the chemistry of sulphur by discussion about those compounds of sulphur they have already met – sulphates, sulphites, sulphur dioxide, and hydrogen sulphide amongst others. The oxidation numbers of sulphur in these compounds should be calculated and the students can be referred to the table of all the oxidation numbers of sulphur given in the *Students' book*.

Students should be referred back to the beginning of this topic. It may be seen that 'expansion of the octet' can occur with sulphur but not with oxygen. The reason for six electrons being available for bonding should be discussed. It should be pointed out that sulphur has an octet of electrons in H_2S but in SO_3^{2-} this is expanded to ten, and in SO_4^{2-} to twelve electrons. The oxidation numbers of ions containing S—S bonds should be discussed, and also the situation in persulphate.

This case is interesting. The persulphate ion is $S_2O_8^{2-}$. If all the oxygen is taken as having oxidation number -2, then the oxidation number of the sulphur must be -7. However, if we look at the structure of the persulphate ion (that is, the relative positions of the various atoms in the ion, ignoring the precise details of their bonding)

```
    O           O
    |           |
O—S—O—O—S—O
    |           |
    O           O
```

two of the oxygen atoms appear as in peroxides, in which they are given oxidation number -1. If this is the case, then the oxidation number of sulphur must be $+6$.

EXPERIMENT 18.5a
Disproportionation in some sulphur compounds

Each student or pair of students will need:

Safety glasses, 1 pair each
Test-tubes, 100 × 16 mm, and test-tube rack
1M hydrochloric acid
Powdered roll sulphur
1M sodium hydroxide solution
0.1M sodium thiosulphate solution

Procedure

Outline instructions are given in the *Students' book*.
The following notes may be useful.

1 In acid solution the sulphur in the sodium thiosulphate disproportionates to give a precipitate of sulphur and sulphur dioxide.

2 Sulphur disproportionates on boiling with 1M sodium hydroxide solution to give sulphide ions and thiosulphate ions. If the solution is acidified the thiosulphate disproportionates as in **1** and sulphur is re-precipitated.

EXPERIMENT 18.5b
Some reactions of sulphite ions, SO_3^{2-}(aq), and of sulphide ions, S^{2-}(aq)

Each student or pair of students will need:

Test-tubes, 100 × 16 mm, and test-tube rack
Access to fume cupboard
0.1M potassium iodide solution
0.1M potassium manganate(VII) solution, acidified with dilute sulphuric acid
0.1M sodium sulphite solution

Optional:

Sodium sulphide solution
1M HCl
Lead(II) nitrate ⎱ solution
  or ethanoate ⎰
Filter paper
Hydrogen peroxide

Procedure

Warn students that sulphur dioxide has a strong choking smell and is especially dangerous to anyone suffering from a respiratory complaint. They should do this reaction on very small quantities in a fume cupboard.

If students carry out the hydrogen sulphide reaction, this MUST be done in a fume cupboard as hydrogen sulphide is extremely poisonous.

Details of the procedure are given in the *Students' book*.
The following notes may be useful.

1 Sulphite ions are oxidized to sulphate ions by manganate(VII) ions in acid solution.
2 Sulphite ions in acid solution reduce iodine to iodide ions.

This section of the *Students' book* ends with an account of the manufacture of sulphuric acid.

Supporting homework

Reading the account of the manufacture of sulphuric acid.

Summary

At the end of this section students should:
1 know some of the reactions of sulphur and its compounds;
2 be aware of the commercial importance of sulphuric acid and understand the application of kinetic and equilibrium concepts to the manufacture of sulphuric acid.

ANSWERS TO PROBLEMS IN THE *STUDENTS' BOOK*

(A suggested mark allocation is given in brackets after each answer.)

1a Acidified potassium dichromate(VI) solution should oxidize tin(II) ions to tin(IV) ions. (3)
 b No reaction is likely. (1)
 c Boron trichloride is hydrolysed to boric acid and hydrochloric acid. (3)
 d Germanium(IV) chloride is hydrolysed to hydrated germanium(IV) oxide and hydrochloric acid. (3)
 Total 10 marks

2a i Iron(III) oxide, Fe_2O_3. ii Aluminium oxide, Al_2O_3. (2)
 b In the case of iron, the silicon dioxide is removed by adding limestone which reacts to form calcium silicate (slag). Treatment of aluminium oxide ore with sodium hydroxide and bubbling in carbon dioxide precipitates aluminium hydroxide but not silicon dioxide. (4)
 c Approximately 1800 K. (2)
 d Aluminium is too reactive to be deposited on the cathode by the electrolysis of an aqueous solution. (2)
 e Amongst the reasons why aluminium is more expensive than iron are that the ore is rarer, it requires more pre-treatment, and electricity is very expensive. (3)
 Total 13 marks

3 Mark by impression. **Total 15 marks**

4 BeO   amphoteric
   B_2O_3 acidic
   GeO_2 acidic
   SO_3 acidic
   Tl_2O_3 basic

The reasons that students give for their choice should be based on the positions that the various elements occupy in the Periodic Table. For the further information of the teacher, equations for some reactions illustrating the nature of these oxides are as follows:

$$BeO + 2HCl \longrightarrow BeCl_2 + H_2O$$

$$BeO + 2NaOH \longrightarrow Na_2BeO_2 + H_2O$$

$$B_2O_3 + 2NaOH \longrightarrow 2NaBO_2 + H_2O$$

$$GeO_2 + 2NaOH \longrightarrow Na_2GeO_3 + H_2O$$

$$SO_3 + 2NaOH \longrightarrow Na_2SO_4 + H_2O$$

$$Tl_2O_3 + 2NaOH \longrightarrow 2NaTlO_2 + H_2O$$

Total 15 marks

5a It has a molecular structure (low melting point and boiling point). (2)
 b It would hydrolyse to give hydrated tin(IV) oxide and hydrobromic acid. (3)
 c Trichloroethane (or suitable alternative). (1)
 d Theoretical yield = 10.7 g.
   Percentage yield = 79.4%. (4)
 e The reaction would be very vigorous (it is exothermic and needs cooling even in solution). (2)
 f The tin(IV) state can be obtained by using bromine as an oxidizing agent whereas lead(IV) cannot be obtained in this way; as you go down the group the element becomes more difficult to obtain in the +4 oxidation state. (2)

Total 14 marks

6a $SO_3^{2-}(aq) + S(s) \longrightarrow S_2O_3^{2-}(aq)$ (2)
 bi Disproportionation occurs giving a sulphur precipitate:

$$S_2O_3^{2-}(aq) + 2H^+(aq) \longrightarrow S(s) + SO_2(aq) + H_2O(l)$$ (3)

ii The iodine is reduced to iodide ions; the colour of the iodine disappears.

$$2S_2O_3^{2-}(aq) + I_2(aq) \longrightarrow S_4O_6^{2-}(aq) + 2I^-(aq) \tag{3}$$

Total 8 marks

7a According to the electrode potential chart in figure 18.18 in the *Students' book*, the oxidation of copper to copper(II) ions has an electrode potential of $+0.34$ V. The reduction of nitrate ions to each of the three principal oxides of nitrogen has an electrode potential which is more positive than $+0.34$ V. (3)

b i It would be expected that concentrated nitric acid would give dinitrogen tetroxide;
ii 7M nitric acid would give nitrogen monoxide and/or dinitrogen monoxide;
iii 2M nitric acid would give nitrogen and/or ammonium ions. (3)
c Nitrogen dioxide can be removed using aqueous sodium hydroxide because it is an acidic oxide, whereas dinitrogen oxide, N_2O, is neutral. (3)
d Nitrous acid disproportionates readily. (1)

Total 10 marks

Appendix 1
Nomenclature, units, and abbreviations

Nomenclature

The naming of chemicals in this course follows the recommendations of the Association for Science Education, which are published in *Chemical nomenclature, symbols, and terminology* (third edition 1984). This book can be obtained from the Association for Science Education, College Lane, Hatfield, Hertfordshire, and all teachers of chemistry at this level should have access to a copy.

Although a knowledge of modern systematic nomenclature is sufficient to complete an A-level course in chemistry with success and with a thorough understanding of the subject, students must realize that many older names remain in current use in industry, commerce, and some areas of higher education. Furthermore, anyone wishing to consult older textbooks, and the chemical literature in general, will have to be conversant with older names. Teachers thus have a duty to acquaint their students with some of the more widely-used of the older, non-systematic names, such as ferrous sulphate, or acetic acid, to quote two examples. For this reason some older names are given in brackets after the systematic names from time to time in both the *Students' book* and the *Teachers' guide*.

The most important aspect of this, of course, must be safety. Nothing should be done which could possibly endanger the health or safety of a student because of a confusion over names, and particular care should be taken to see that any such risk is avoided.

Units

The internationally accepted system of units known as the *Système International d'Unités*, or SI units, is followed in all Nuffield Advanced Science Publications. This is a coherent system based on the seven units of metre (symbol m), kilogramme (kg), second (s), ampere (A), kelvin (K), candela (cd), and mole (mol).

Definitions of these units, and a list of units that are derived from these seven basic units, are given in the *Book of data*. Some further comments however are worth making here.

Multiples of units – Powers of ten are indicated by prefixes to the names of the units: for example, a thousand metres is known as a kilometre (km). Preference is given to intervals of 10^3; thus for *preference* a length should be quoted as 125 millimetres (mm) rather than 12.5 centimetres (cm). Centimetres are not, however, forbidden, and may be used if desired.

The nanometre – When recording measurements such as the wavelength of light, or interatomic distances, a widely used unit has been the Ångstrom, which is 10^{-10} m. The preferred SI unit is 10^{-9} m, known as a nanometre (nm). In this course, such distances are quoted in nanometres: 1 nm = 10 Å.

The cubic decimetre and cubic centimetre – The cubic metre (m^3) is an inconveniently large unit of volume for practical purposes, and chemists will normally use the cubic decimetre (dm^3) and the cubic centimetre (cm^3). The word litre is now taken as a special name for the cubic decimetre. Although this name may be used in everyday speech it is an unnecessary extra word, and neither it nor the 'millilitre' are used in this course.

The kelvin – The unit of thermodynamic temperature is the kelvin (*not* 'degree kelvin'), symbol K. For practical purposes the texts often refer to temperatures in degrees Celsius (centigrade), symbol °C. 0 °C = 273 K.

Pressure – The unit of energy is the joule, J. Force is expressed in joules per metre, $J\ m^{-1}$, or *newtons*, N. Pressure is expressed in newtons per square metre, $N\ m^{-2}$, or *pascals*, Pa. For practical purposes it is usual to talk about atmospheres, and to measure pressures in millimetres of mercury, mmHg, and all three units are used at various places in this book.

$$1 \text{ atmosphere} = 760 \text{ mmHg} = 101\,325\ N\ m^{-2} = 101.325\ kPa$$

Wavenumbers – Infra-red absorption spectra have percentage transmittance plotted against wavenumber, units cm^{-1}. Here, we have kept to the standard practice of spectroscopists, rather than using wavelength. The wavenumber is the reciprocal of the wavelength.

Abbreviations

The principal abbreviations or symbols for units have already been mentioned; others are explained in the text, and all are given in the *Book of data*. State symbols are used in many equations; these have the following meanings:

(s) = solid (g) = gas
(l) = liquid (aq) = dissolved in water

Other abbreviations, signs, and symbols follow the recommendations given in *SI units, signs, symbols, and abbreviations* (1981), published by the Association for Science Education.

Appendix 2
Apparatus

In this *Teachers' guide* the apparatus required to perform each experiment is listed before the experiment is described. Some experiments require apparatus that is not generally available, and for others it is possible to construct apparatus as an alternative to that which is commercially available. In these cases, reference is made to this Appendix, and the construction of this apparatus is described here.

1   Syringes and accessories

Gas syringes – All-glass gas syringes should be used for all quantitative work with gases, and not the plastic variety. Plastic syringes are not sufficiently free-running, and cannot be heated as required for experiment 3.4b. Suitable all-glass gas syringes of 100 cm^3 capacity, graduated in intervals of 1 cm^3, can be obtained from most of the usual laboratory suppliers.

Hypodermic syringes and needles – Many types of hypodermic syringes are available. For experiment 3.4b those of 2 cm^3 capacity are suitable; they should be two-piece and all glass, with 'luer' fitting. They will require hypodermic needles of the same 'luer' fitting; these are usually sold as separate items. Care should be taken to see that the fittings are the same; 'record' fitting needles are not suitable for 'luer' syringes.

Caps – Rubber caps for the 100 cm^3 gas syringes can be obtained from the usual suppliers.

Self-sealing caps for the hypodermic needles can be obtained by using silicone rubber bungs obtainable from ESCO (Rubber) Limited, Sterilin House, Clock House Lane, Feltham, Middlesex. A suitable size is E5.

Steam jackets – Various steam jackets are available commercially.

2   Direct-vision spectroscope (Topics 2 and 4)

A direct-vision spectroscope can be made out of any hollow tube of a suitable size. One end of the tube is sealed except for a narrow slit, and a replica diffraction grating is put over the other end.

Materials required
Tube, cardboard or metal, about 12 cm long and 2–4 cm diameter
Matt black paint (aerosol, for preference)

Aluminium foil
Two rubber bands
Diffraction grating replica. Available from Griffin & George, ref. no. XFY–530–B.
Optional: second tube, just fitting over the first; two lengths of about 3–4 cm.

Construction

The inside of the long tube should be sprayed with matt black paint to prevent internal reflections.

One end of the tube is covered with aluminium foil, which is folded over the end of the tube and held in place with a rubber band. A slit about 1 cm long is then cut in the foil with a razor blade, so that the slit is centrally placed across the end of the tube.

The other end of the tube is covered with aluminium foil in a similar manner, but this foil has a 5 mm diameter hole cut in it using a cork borer, and a piece of replica diffraction grating stuck over the hole. The foil should be positioned carefully so that the hole is centrally placed over the end of the tube, and the grating lines are parallel with the slit in the foil at the other end of the tube.

The two ends of the assembly are easily damaged. They can be protected by fitting a piece of tubing of slightly larger diameter over both ends of the first tube, so as to project about 1 cm (see figure A1).

In use, the tube is held with the slit pointing at the source of light, and the grating close to the eye. The direct slit will be seen straight through the grating, with the first order spectrum on one side.

Figure A1
Direct-vision spectroscopes.
Photograph, B. J. Stokes.

3 The use of radio valves to find the ionization energy of a noble gas (Topic 4)

At one time a number of different radio valves could be used in experiments to determine the first ionization energies of the gases used to fill the valves. By a suitable choice of valve, and of electric circuit, the ionization energies of each of the noble gases, and of some other elements including hydrogen and mercury, could be determined (see 'Measurement of ionization potentials of the rare gases', B. E. Dineen and R. S. Nyholm, Journal of the Royal Institute of Chemistry, 1963, volume 87, pages 110–15). Most of these valves are no longer manufactured, but two can still be obtained, the 884 containing argon, and the EN91 containing xenon.

A simple circuit is shown in *Students' book I*, page 81, and its principle is explained there. In trials, using 884 valves in this circuit, a sudden jump in the grid current and a visible argon discharge occurred at an applied voltage of 13.5 V (although the expected value for argon is 15.75 V). This low value may be due to the spacing of the electrodes, the gas pressure, or some other factor, but it can still be used to illustrate the principle of the method. If this circuit is used, a 100 ohm resistance should be put in series with the grid so as to limit the grid current after ionization; otherwise the valve may burn out. The anode (pin 3) should be connected to the grid (pin 5), effectively converting the valve into a diode.

The circuit can be set up by mounting an octal base and a wire-wound potentiometer on a base board, and wiring them to 4 mm sockets fixed to the same board (see figure A2). On the board illustrated, the circuit diagram has been drawn on top of the board; the actual wiring is carried underneath.

Materials required
Octal base
500 ohm 3 W wire-wound potentiometer
2 yellow 4 mm sockets
3 black 4 mm sockets
3 red 4 mm sockets
These items can be obtained from RS Components Limited, 13–17 Epworth Street, London EC2P 2HA.
Wood for baseboard, approximately 15 cm × 30 cm
Fixing screws
Connecting wire
Use of soldering iron, to make electrical connections, and drill for making holes in the baseboard

The following must then be plugged into the board
Source of 20 V d.c.
Source of 6.3 V a.c.
High resistance voltmeter, 0–20 V
Milliammeter, 0–50 mA

a

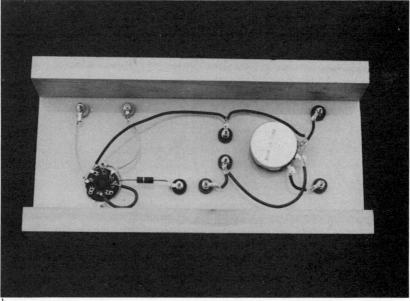

b

Figure A2
Circuit board for the determination of the ionization energy of argon.
a top view
b view from underneath.
Photographs, B. J. Stokes.

Valve, type 884 (for source of these valves, see below)
Connecting wires, fitted with 4 mm plugs, will be required

A more complicated circuit is given in *Teachers' guide I*, pages 62–3. The various voltages and currents given there should be taken as a general guide only, as values have been found to vary between valves from different sources. Great care must be taken when setting up the circuit and writing suitable instructions for students, as valves are easily burned out if too great a current is allowed to flow. It is very convenient to have the complete circuit mounted ready for use. As an example, figure A3 shows a self-contained apparatus made by C. J. V. Rintoul, which is ready for use when plugged into the mains electricity supply. The circuit is constructed of wire having white insulation, and held under a perspex cover. To make it easy for the student to follow, the wiring is set out in the same way as the circuit diagram on page 63 of *Teachers' guide I*. The mains lead and the valve are stored in a covered recess in the base of the box when not in use.

Figure A3
Self-contained apparatus for the determination of the ionization energy of argon.
Photograph, B. J. Stokes.

Materials required

Transformer and rectifier arranged to supply
 6.3 V a.c.
 20 V d.c.
 3 V d.c.

Mains switch
Pilot light
Fuse
250 ohm, 3 W wire-wound variable resistor
Octal base
High-resistance voltmeter, 0–20 V
Microammeter, 0–50 A
Wooden box with perspex lid
Connecting wire
Fixing screws
Valve, type 884

Valves can be obtained from Zaerix Electronics Limited, Electron House, Cray Avenue, St Mary Cray, Orpington, Kent BR5 3PN.

4 Apparatus to demonstrate the principle of the mass spectrometer (Topic 4)

Several different models which demonstrate this principle can be constructed. One of them is described here.

The model consists of a tray in which there are three compartments, as shown in figure A4. Ball bearings of different masses are rolled down a glass tube into the tray and are then deflected by a magnet. Adjustment of the slope of the glass tube and of the position of the magnet enable the ball bearings to be deflected into the three compartments according to their masses.

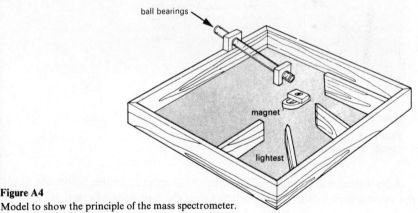

Figure A4
Model to show the principle of the mass spectrometer.

Materials required
1 Eclipse horseshoe magnet – width across limbs 44 mm, thickness 10 mm
Ball bearings diameters 11 mm, 6 mm, and 4 mm
Wood for tray
Glass tube about 30 cm long, internal diameter greater than 11 mm

The dimensions of the tray are not critical. Nominal dimensions are given in figure A5.

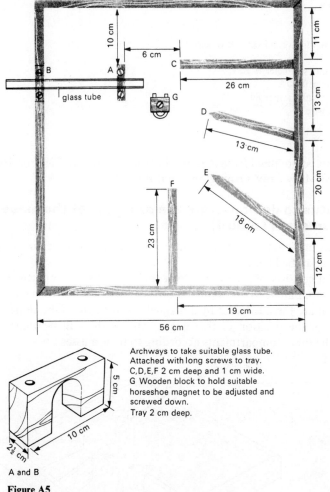

Figure A5
Details of the model.

Adjustment will be required, moving the magnet and tube, before the model works satisfactorily. During this adjustment, the ball bearings that stick to the magnet will become magnetized and should be dropped several times to demagnetize them.

5 Electrical compensation calorimeter (Topic 6)

An electrical compensation calorimeter consists of a vacuum flask, in which the chemical reaction is carried out, an electric immersion heater which can be placed inside the flask, and a thermometer which can be read to the nearest 0.1 °C.

Materials required
Vacuum flask, 500 cm³ ('Thermos flask')
Lamp, 12 V 20–24 W
Thermometer, 0–50 °C in 0.1 °C intervals
Cork to fit flask
Leads
Araldite
A soldering iron will be required

A standard size vacuum flask of about 500 cm³ capacity, such as is sold for keeping drinks warm, is suitable for this apparatus. These flasks are protected by a plastic case, and stand upright on a flat base. Unprotected vacuum flasks should *not* be used for experimental work; if they break there is a considerable risk of injury from flying glass.

The heater can be made from a 12 V 24 W lamp. The lamp will be immersed in the solution in the flask, and so the electrical connections to it must be waterproof; an ordinary lampholder will not do. The electric leads to the lamp must be soldered to the lamp terminals, and the whole of the end of the lamp and leads completely covered with Araldite to protect them. This operation must be very carefully done if it is to be successful.

The flask should be fitted with a cork, drilled to hold the thermometer and the leads to the lamp.

A diagram of the apparatus is given in *Students' book I*, page 145; a photograph of a typical heater assembly is given here as figure A6.

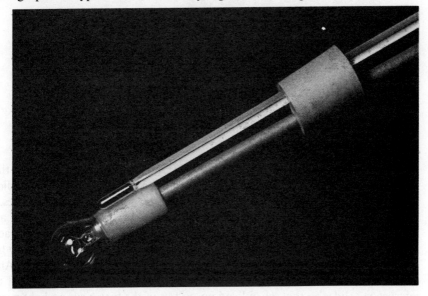

Figure A6
Heater for the electrical compensation calorimeter.
Photograph, B. J. Stokes.

6 Crossed polaroid assembly (Topic 7)

Details of the construction of a simple assembly are given in figure A7.

Materials required
4 microscope slides
Polaroid sheet
Wood
Quick-drying adhesive

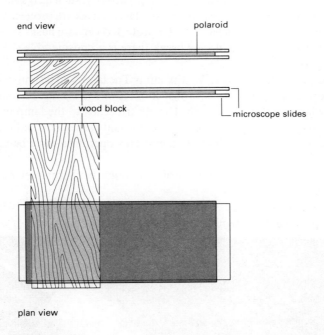

Figure A7
Crossed polaroid assembly.

A piece of polaroid sheet is cut to the size of the microscope slides. It is then placed between two slides, to protect it from dust and scratching, and the slides are stuck together by application of a quick-drying adhesive round the edges.

When set, it should be supported horizontally above a torch bulb and holder. With the light on, the remainder of the polaroid should be placed horizontally above the microscope slide and rotated until the light bulb, when viewed through the two pieces of polaroid, appears to be extinguished. After careful marking, a second piece of polaroid can now be cut to the size of the microscope slides so that when lined up on top of the first piece no light passes through the two. The second piece is then fixed between two microscope slides in the same way as the first.

The two polaroid sandwiches are then stuck to a piece of wood so that they are held parallel to one another, about 2 cm apart, as shown in figure A7. Before the adhesive is set, the extinction of the light should be checked, and any necessary adjustment to the alignment made.

In use, the wood is held in a clamp stand so that the polaroid sheets are horizontal, and a torch bulb in a holder is placed beneath. Alternatively, a wooden stand can be made as shown in figure A8.

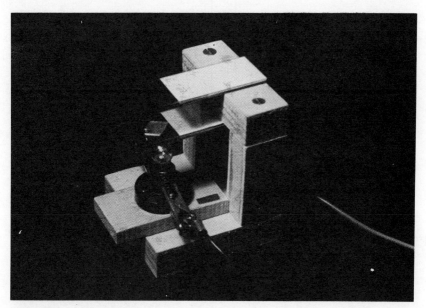

Figure A8
Crossed polaroid assembly.
Photograph, B. J. Stokes.

7 Polarimeter (Topic 11)

Details of the construction of a simple polarimeter, devised by K. Frazer, are given in figures A9 and A10.

Materials required
2 pieces of polaroid, 1.5 cm square
Specimen tube, 150 × 25 mm
3 Terry clips
Wood for mount, 18 × 2.5 × 2.5 cm
Protractor, 0–180°, 10 cm diameter
Sewing needle for pointer
2 corks
Paraffin wax
6 V bulb and holder

204   Appendix 2

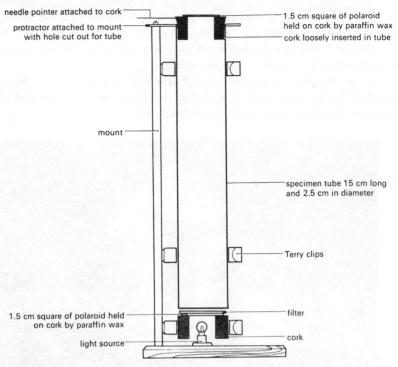

Figure A9
Home-made polarimeter, side view.

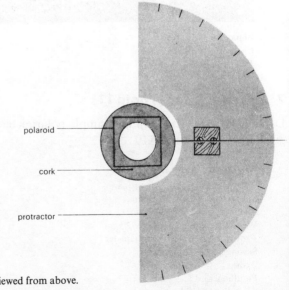

Figure A10
Home-made polarimeter, viewed from above.

The analyser is mounted on the cork which is loosely inserted into the specimen tube. The polarizer is mounted below the specimen tube.

For good, clear results it is important that the base of the specimen tube is absolutely flat, or it will distort the light passing through it. Suitable tubes can be bought, or a substitute made by sticking a flat piece of window glass across one end of a piece of glass tubing using a suitable adhesive. If this is done, care is needed to get the end of the glass tubing quite flat so that it makes good contact with the glass to be stuck to it. Attention must also be given to the edges of the flat base so that they do not cut anyone picking them up. They should be ground smooth, or covered with strips of insulating tape.

8 Colorimeter (Topics 14 and 16)

Simple colorimeters suitable for work at this level are available from the principal laboratory equipment suppliers. These instruments afford a quick and easy method of determining the concentrations of solutions which are coloured. Basically they consist of the components shown in figure A11.

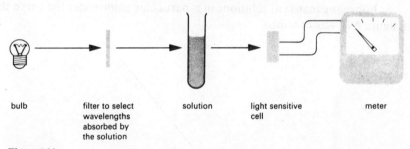

bulb   filter to select wavelengths absorbed by the solution   solution   light sensitive cell   meter

Figure A11
Colorimeter.

The light-sensitive cell may either be a selenium cell which produces an e.m.f. proportional to the intensity of the light falling on it, or may be a cadmium sulphide cell, the electrical resistance of which is proportional to the intensity of the light falling on it. Either way the meter reading gives an indication of the intensity of light emerging from the solution.

The connection between the intensity of light emerging from the solution and the molarity of the absorbing species in the solution is given by:

$$\lg\left(\frac{I_0}{I}\right) = \lg\left(\frac{m_0}{m}\right) = \varepsilon l M$$

or $\lg\left(\frac{I_0}{I}\right) = \lg\left(\frac{m_0}{m}\right) \propto M$

where I_0 = intensity of incident, *monochromatic* light giving meter reading m_0

I = intensity of emergent light giving meter reading m
M = molarity of solution
l = path length of light through solution
ε = molecular extinction coefficient

The colorimeter is normally used by first adjusting the meter reading to maximum for I_0 by inserting a tube of pure solvent, and then altering the intensity of light by adjusting a shutter placed between the bulb and the cell (or by altering the current flowing through the bulb). The tube of solution is then put into the colorimeter and a meter reading proportional to I determined. It is *most important* that this procedure is adopted for every reading if possible, unless the electricity supply is very stable.

Unfortunately, the law quoted above is not obeyed accurately by all solutions. Also the meter reading may not be accurately proportional to the intensity of light, so the instrument must be calibrated before use. This will also indicate for what range of concentration of a particular substance the colorimeter can be used.

For manganate(VII) solutions in a particular colorimeter the curve shown in figure A12 was obtained.

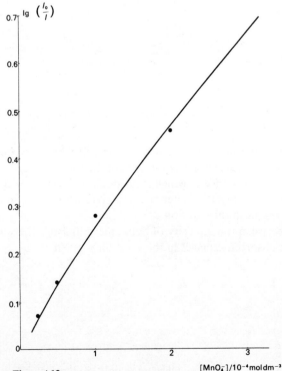

Figure A12
Calibration curve for the colorimeter.

If a calibration curve is being used, it is easier to plot $\left(\dfrac{I}{I_0}\right) = \left(\dfrac{m}{m_0}\right)$ than $\lg\left(\dfrac{I_0}{I}\right) = \lg\left(\dfrac{m_0}{m}\right)$, as m_0 is normally 1, 10, or 100 or some number which is a useful number to have as a denominator.

Such a graph (figure A13) indicates that the colorimeter will not measure concentrations accurately above about 0.0003M.

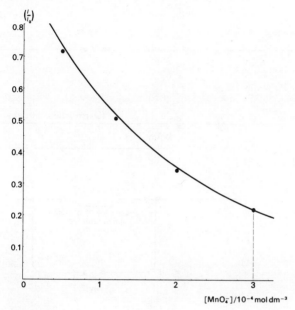

Figure A13
Alternative form of calibration curve.

Before these curves can be constructed, the most suitable filter must be chosen. This should select that band of wavelengths of light which are most strongly absorbed by the solution. This is satisfied by the filter which gives the lowest value of $\left(\dfrac{I}{I_0}\right)$ for a particular solution.

Procedure

It is convenient to supply students with a calibration curve for a particular filter, colorimeter, and solution. They will require a test-tube for the pure solvent and an optically matched one for the solution. To take a reading, the meter is adjusted to maximum with pure solvent in place, and then the reading is obtained with solution in place. It is best then to check again with solvent in place and obtain a meter reading with solution in place once again. The value of $\left(\dfrac{I}{I_0}\right)$ so obtained can then be turned into a molarity using the calibration chart.

9 Potentiometers for e.m.f. measurements (Topic 15)

As an alternative to a solid-state or other high resistance voltmeter, a potentiometer can be used to measure a potential difference accurately. The circuit is shown in figure A14.

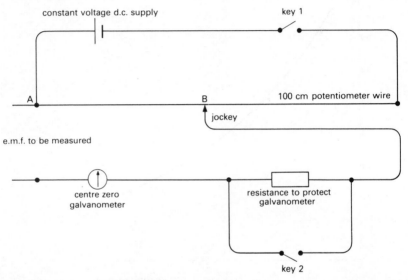

Figure A14

Apparatus for this circuit can be bought, or borrowed from the Physics Department.

For each potentiometer the following materials will be needed
2 keys for making and breaking the circuit
1 straight wire potentiometer, 100 cm^3, 0.05 Ω cm^{-1} with jockey
1 centre zero galvanometer (sensitivity at least 10 µA per division)
1 protective resistance for galvanometer, approximately 1000 Ω
1 2 V accumulator
Leads
Standard cell, Weston or mercury battery

This method depends on opposing the current flowing in a circuit, of which the cell is a part, by a potential difference from another source of electricity, and adjusting this second p.d. until no current flows. The counter-voltage thus applied is equal to the e.m.f. of the cell being studied.

Procedure

With key 1 closed, the position of the jockey (a sliding contact) is adjusted until the galvanometer reading is zero. This gives an approximate balance point only, because of the protective resistance in series with the galvanometer. A more accurate balance point can be obtained by shorting out the protective resistance (close key 2) and re-adjusting the sliding contact.

The fall in potential around the top part of the circuit in figure A14 occurs mainly along the potentiometer wire. When no higher accuracy is required, the e.m.f. of the cell being tested can be calculated easily from the known voltage of the d.c. supply. Thus, a 2 V accumulator could be used as the source of this current. The potential fall along the potentiometer wire is $\frac{2}{100} = 0.02$ V per cm. If the two parts of the circuit are in balance when the distance AB is 37.0 cm, the e.m.f. of the cell being tested is $37.0 \times 0.02 = 0.74$ V.

To obtain a more accurate value, the potentiometer wire must be calibrated more carefully. This is done by replacing the test cell by a standard cell, the e.m.f. of which is accurately known for the temperature at which the measurement is being made. For example, a Weston standard cell* might be used which has an e.m.f. of 1.018 V at 20 °C. If, for this cell, the balance point is such that the distance AB is 49.4 cm, the e.m.f. of the cell is $\frac{37.0 \times 1.018}{49.4} = 0.726$ V.

10 Hydrogen electrode (Topic 15)

A number of quite simple versions of this electrode can easily be constructed. Three possible designs are shown in figure A15.

Expense is saved if copper wire is soldered to a short length of platinum wire (use zinc chloride/hydrochloric acid flux and ordinary soft solder).

The platinum wire must be covered with electrolytically-deposited platinum over the lower inch or so. This is done by using another platinum wire as a second electrode, and immersing both in a 3% solution of platinum(IV) chloride containing a small proportion (approximately 0.05%) of lead ethanoate, connecting the electrodes to a d.c. supply (about 4 V), and reversing the current every half minute for 15 minutes. The electrodes are then washed with distilled water, and occluded chlorine is removed by placing the electrodes in sodium ethanoate solution (approximately M) and again passing current for 15 minutes, reversing every half minute. The electrodes are then washed with distilled water and stored in distilled water when not in use.

A cylinder is the most convenient supply of hydrogen gas but must be fitted with a reducing valve, in order to provide a very slow stream of gas. Alternatively,

---

* The cell diagram for the Weston cell is
Cd(s) | CdSO$_4$(aq,satd.) ⋮ Hg$_2$SO$_4$(aq,satd.) | Hg(l); $E_{293} = +1.018$ V.

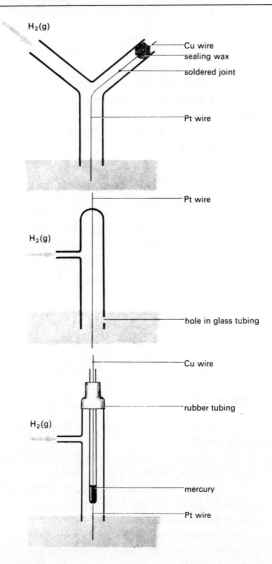

Figure A15
Types of hydrogen electrode.

a filter flask can be used as an aspirator (see figure A16). This can be filled from a cylinder or from a suitable gas generating apparatus (in this case put dilute manganate(VII) solution and a little dilute acid in the aspirator flask to remove impurities liable to poison the platinum electrode).

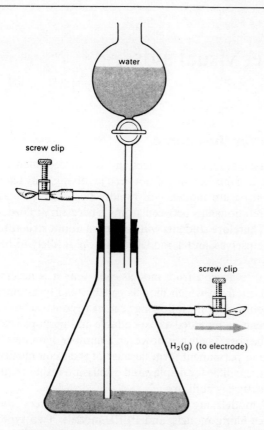

Figure A16
Aspirator for a hydrogen electrode.

Appendix 3
Models and other visual aids

Structure models for the course

In any serious attempt to assist students to understand the three-dimensional structures of substances the use of models should not need justifying. Any teacher who attempts to construct structure models will rapidly find that a detailed understanding of the spatial relationships between atoms is necessary in order to complete them successfully. Therefore students will best learn about structures if they construct models for themselves, and a 'model workshop' is likely to be an annual feature of this chemistry course.

Given that the objective is to construct models each year it is necessary either to have component parts from which models can be readily assembled when required and then demounted for the following year, or to construct models from cheap materials which can be regarded as expendable and perhaps sold to their constructors to defray the cost. There are, however, a number of models that the teacher will wish to have in permanent form because of their complexity or because their main teaching use will be for display and discussion rather than for consideration of how the structure is built up.

Three main classes of models are referred to in this *Teachers' guide*: ball-and-spring models, space-filling models, and PEEL models. Two types of space-filling models are used: tangential contact models and Stuart-type models. These different models are used so that different structural features can be illustrated to greatest advantage.

Ball-and spring models

Ball-and-spring models have the individual atoms widely separated, and are particularly useful in small molecules to show bond angles, and in large structures to enable the positions of atoms in the body of the structure to be seen clearly.

A good supply of components from which ball-and-spring models can be made should be available for student use. Those of the Linnell type are particularly suitable. The plastic spheres representing atoms are available in a number of colours, and are drilled with holes at various angles. The colour code is as follows:

hydrogen: white   phosphorus: brown
carbon: black   sulphur: yellow
nitrogen: blue   halogens: green
oxygen: red   metals: yellow

The holes are drilled as follows.
4 coordinated: 4 holes all at 109° to each other
5-hole trigonal pyramid: 3 holes in one plane at 120° to each other, and 2 holes perpendicular to this plane
6 coordinated: 6 holes at 90° to each other
8 coordinated: 8 holes pointing from the centre to the 8 corners of a cube.

The following ball-and-spring models are listed in the *Teachers' guides*.

Elements: P_4, S_8, Cl_2 (*Teachers' guide I*, pages 8 and 150). Models of these molecules are illustrated on overhead projection transparency number 69. Requirements for each model are:

P_4 4 phosphorus atoms (brown, 4 coordinated)
       6 springs, about 6 cm length

S_8 8 sulphur atoms (yellow, 4 coordinated)
       8 springs, about 6 cm length

Cl_2 2 chlorine atoms (green, any coordination)
       1 spring, about 6 cm length

Compounds: CH_4 (*Teachers' guide I*, pages 136 and 158); P_4O_{10}, $SiCl_4$, PCl_5 (*Teachers' guide I*, page 150); $BeCl_2$, BF_3, SiH_4, C_2H_6, NH_3, PH_3, H_2O, H_2S, HF, HCl, CH_2=CH_2, HC≡CH, CO_2, HCN, 4-methoxybenzoic acid (*Teachers' guide I*, page 158). Models of these molecules are illustrated as indicated in the following lists of requirements for each model.

The following models are illustrated on overhead projector transparency number 70:

$SiCl_4$ 1 silicon atom (black, 4 coordinated)
       4 chlorine atoms (green, any coordination)
       4 springs, about 6 cm length

PCl_5 1 phosphorus atom (brown, 5-hole trigonal bipyramid)
       5 chlorine atoms (green, any coordination)
       5 springs, about 6 cm length

The following model is illustrated on overhead projector transparency number 71:

P_4O_{10} 4 phosphorus atoms (brown, 4 coordinated)
       10 oxygen atoms (red, 4 coordinated)
       16 springs, about 6 cm length

The following models are illustrated on overhead projector transparency number 76:

CH$_4$	1 carbon atom (black, 4 coordinated) 4 hydrogen atoms (white, 1 hole) 4 springs, about 6 cm length
NH$_3$	1 nitrogen atom (blue, 4 coordinated) 3 hydrogen atoms (white, 1 hole) 3 springs, about 6 cm length
H$_2$O	1 oxygen atom (red, 4 coordinated) 2 hydrogen atoms (white, 1 hole) 2 springs, about 6 cm length
HF	1 fluorine atom (green, 4 coordinated) 1 hydrogen atom (white, 1 hole) 1 spring, about 6 cm length

The following models are made in a similar manner to a model already described:

SiH$_4$	as CH$_4$
PH$_3$	as NH$_3$ but with a phosphorus (brown) atom in place of a nitrogen (blue) atom
H$_2$S	as H$_2$O but with a sulphur (yellow) atom in place of an oxygen (red) atom
HCl	as HF

The following models are made as is now described:

BeCl$_2$	This is a linear molecule, Cl—Be—Cl. Requirements are: 1 beryllium atom (use any 6 coordinated atom) 2 chlorine atoms (green, any coordination) 2 springs, about 6 cm length
BF$_3$	This molecule has all three bonds in one plane, at 120° to each other. Requirements are: 1 boron atom (use any 5-hole trigonal bipyramid) 3 fluorine atoms (green, any coordination) 3 springs, about 6 cm length
C$_2$H$_6$	The molecule of ethane is made in a similar manner to that of methane, CH$_4$. Requirements are: 2 carbon atoms (black, 4 coordinated) 6 hydrogen atoms (white, 1 hole) 7 springs, about 6 cm length

Models of the following molecules are illustrated in figure A17:

CH$_2$=CH$_2$	2 carbon atoms (black, 4 coordinated)

	4 hydrogen atoms (white, 1 hole)
	6 springs, about 6 cm length
HC≡CH	2 carbon atoms (black, 4 coordinated)
	2 hydrogen atoms (white, 1 hole)
	5 springs, about 6 cm length
O=C=O	1 carbon atom (black, 4 coordinated)
	2 oxygen atoms (red, 4 coordinated)
	4 springs, about 6 cm length
H—C≡N	1 hydrogen atom (white, 1 hole)
	1 carbon atom (black, 4 coordinated)
	1 nitrogen atom (blue, 4 coordinated)

4-methoxybenzoic acid
  2 carbon atoms (black, 4 coordinated); these are used for the methoxy- and carboxylic acid groups
  6 carbon atoms (black, 5-hole trigonal bipyramid); these are used for the benzene ring
  3 oxygen atoms (red, 4 coordinated)
  8 hydrogen atoms (white, 1 hole)
  20 springs, about 6 cm length

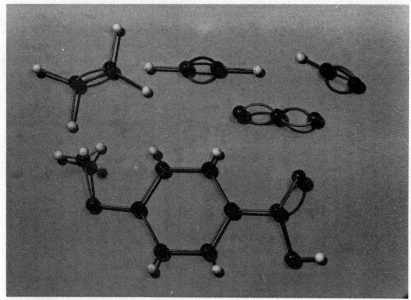

Figure A17
Some ball-and-spring models.
Photograph, B. J. Stokes.

Larger structures: diamond, graphite, sodium chloride, caesium chloride, calcium fluoride, and zinc sulphide (*Teachers' guide I*, page 140 and following pages).

Diamond and graphite: illustrations of the diamond and graphite structures are given on overhead projector transparency number 65. The use of small springs is recommended in order to keep these models fairly rigid. Requirements are:

Diamond
    30 carbon atoms (black, 4 coordinated)
    30 springs, about 3 cm length
    This number is sufficient for four atoms along each side of a tetrahedron; more can be added with advantage.

Graphite
    39 carbon atoms (black, 5-hole trigonal bipyramid)
    45 springs, about 3 cm length
    10 springs, about 6 cm length

Illustrations of the remaining four structures are given on overhead projector transparency number 64. Further details of the construction of the structures can be seen from figure A18; requirements are as follows:

Sodium chloride
    14 sodium ions (yellow, 6 coordinated)
    13 chloride ions (green, 6 coordinated)
    54 springs, about 6 cm length

Caesium chloride
    8 caesium ions (yellow, 8 coordinated)
    27 chloride ions (green, 8 coordinated)
    64 springs, about 6 cm length

Calcium fluoride (fluorite)
    14 calcium ions (yellow, 8 coordinated)
    8 fluoride ions (green, 4 coordinated)
    32 springs, about 6 cm length

Zinc sulphide (zinc blende)
    4 zinc ions (brown, 4 coordinated)
    10 sulphide ions (yellow, 4 coordinated)
    16 springs, about 6 cm length

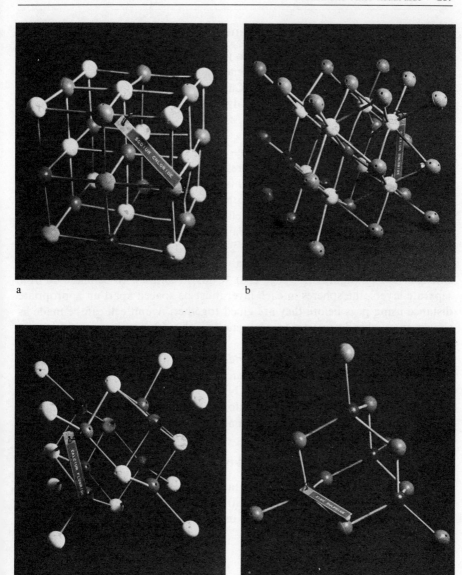

Figure A18
Ball-and-spring models of crystal structures.
a sodium chloride
b caesium chloride
c calcium fluoride
d zinc sulphide.
Photographs, B. J. Stokes.

Space-filling models – tangential contact type

These models are particularly useful to illustrate metal structures. They can be cheaply made from expanded polystyrene foam spheres, glued together in layers or in any other desired arrangement. Care must be taken in the choice of glue; many quick-drying cements dissolve polystyrene foam rapidly, and are therefore unsuitable.

Hexagonal close-packed and face-centred cubic structures These are best illustrated using polystyrene foam spheres glued together in layers, so that several layers can be piled on top of one another in ABAB or ABCA sequences. Diagrams of suitable arrangements are given in *Students' book I* (figures 7.13 and 7.14 on pages 206 and 207). In addition, a unit cell of each structure should be made; these are illustrated on overhead projector transparency number 60.

Body-centred cubic structure Suitable models can be made as illustrated in *Students' book I* (figure 7.15, page 208). If this structure is to be made up from separate layers, the spheres in each layer must be spaced apart an appropriate distance using pegs before they are glued together. A unit cell can be made as shown on overhead projector transparency number 60.

Sodium chloride and caesium chloride structures Tangential-contact models of these structures are referred to in *Teachers' guide I*, page 140. Instructions on how to make them are given in Revised Nuffield Chemistry *Teachers' guide II*, on pages 154, 161, and 162. The sodium chloride space-filling structure is illustrated on overhead projector transparency number 62.

Space-filling models – Stuart type

These models give an excellent representation of the shapes of molecules, and are particularly valuable in organic chemistry. Sets of components can be bought, and should be available for students to use. Some molecular models made from one such set are shown in figure A19.

Models and other visual aids 219

Figure A19
Stuart-type space-filling models.
Photograph, B. J. Stokes.

PEEL models

PEEL models are expanded polystyrene shapes, with inserts which accept the springs used for the Linnell molecular models. The word PEEL stands for 'Probability Envelopes of Electron Location' and the various shapes represent single bond sigma orbitals, protonated orbitals, lone-pair non-bonding orbitals, and π orbitals of various shapes. Illustrations of these models are to be seen in the organic chemistry topics in the *Students' books*; for example in *Students' book I* on pages 234, 238, and 303.

These models are obtainable from Griffin and George Ltd., Ealing Road, Alperton, Wembley, Middlesex HA0 1HJ.

Overhead projection originals

The overhead projector has become such an established part of the teacher's equipment that it needs no introduction here. A set of overhead projection originals was prepared for the first edition of Nuffield Advanced Chemistry, and this set has now been extensively revised (Revised Nuffield Advanced Chemistry *Overhead projection originals*, published by Longman Group Limited, ISBN 0 582 35483 8). The revised set consists of 129 originals, a number of which have additional overlays. The originals are numbered in the order in which they can be of use in the course. They can be used in three ways: as originals from which transparencies can be prepared for use on the overhead projector, as originals from which duplicate copies can be produced for distribution to students, or as small-scale wallcharts to be pinned up in the classroom.

To use the originals in either of the first two ways an infra-red copying machine is required. To make an overhead projection transparency, a sheet of specially prepared acetate film is placed on top of the printed original and the two are passed through the copier. The transparency is produced in a few seconds ready for use.

Thermal stencils are available for both spirit duplicators and those of a Gestetner or Roneo type. These can be passed through an infra-red copier together with the printed original, to provide stencils from which many copies of the original can be made for class use.

Full details of the type of copying material needed, and its method of use, should be obtained from the manufacturers of the infra-red copying machine to be used.

Transparencies have certain advantages over exclusive reliance on a board on which the teacher writes whilst teaching. For example, tables of data can readily be presented for class discussion; infra-red spectra can be examined for characteristic absorption bands by teacher and student together; details of complex apparatus are easily shown to students; and molecular models can be made available as accurate drawings. Furthermore, all of these items are available for repeat presentation at short notice if required.

Transparencies can be used in a variety of ways. Thus the full content need not be revealed at once but portions may be covered with opaque material and exposed only when required. In other cases the content can be on a set of separate transparencies which in use are overlaid on each other in sequence. The transparencies can also be written on, or important portions highlighted with colour, by means of water-based fibre-tipped pens, and afterwards the transparency can be wiped clean for re-use.

The storage of transparencies so that they are readily available for use needs to be properly organized. If each transparency is fixed to a cardboard mounting frame (using a pressure-sensitive tape) they can be numbered on the frame and stored in an ordinary cardboard box in numerical order. Kept in this way a collection of a hundred transparencies can be handled with ease.

Display boards

Teachers may find display boards useful at various points during the teaching of this course.

Energy-level display board (*Teachers' guide I*, page 67, and *II*, page 139) This board has its most important use in Topic 4 when electronic energy levels are first discussed. As well as being an effective method of displaying energy levels and the arrangement of electrons within those levels, the board is also useful because an important concept is introduced by means of an interesting demonstration and thereby rendered more memorable for many students. The display board (figures A20 and A21) is readily constructed from plywood with wooden shelves to represent the energy levels. Beads of 2 cm diameter are used to represent electrons and are fitted over headless nails on the energy levels as required. A box fitted behind the board can be used to store a supply of beads.

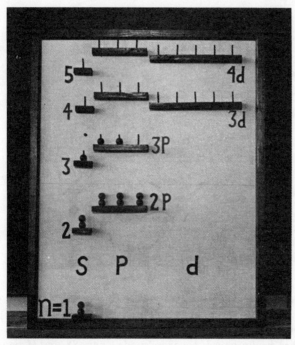

Figure A20
An energy level display board.
Photograph, B. J. Stokes.

Chemical specimens display board (*Teachers' guide I*, page 13) This board is useful to support the study of inorganic substances, so that a wide variety of different compounds can be viewed easily. For example, during the study of Topic 2 it can be used to display samples of the s-block elements and their more important compounds. The board should be constructed from a white plastic-surfaced wood (Contiboard) so that it can be written on with a felt pen. Spring clips are fastened to the board at regular intervals, to hold specimen tubes

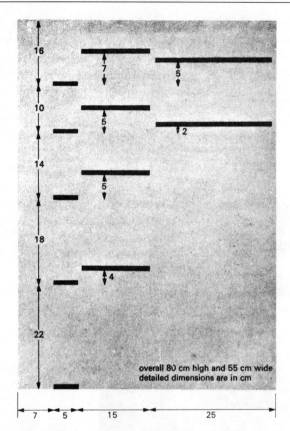

Figure A21
Suitable dimensions for an energy level display board.

containing the materials to be displayed. Suitable dimensions will depend upon the sizes of the spring clips and specimen tubes that are available. An example of such a board is shown in figure A22.

Oxidation number display board (*Teachers' guide I*, page 88, and *II*, page 187) This board can be constructed in a similar manner to the chemical specimens board, with spring clips spaced at suitable intervals, but oxidation numbers are painted down one side. The board is useful in Topic 5, when samples of the elements and their compounds can be arranged to illustrate the oxidation number of the halogen in each case, and again in Topic 16 when the transition elements can be displayed to illustrate their different oxidation numbers. Here, as elsewhere, students should be encouraged to build up the display using their own laboratory material.

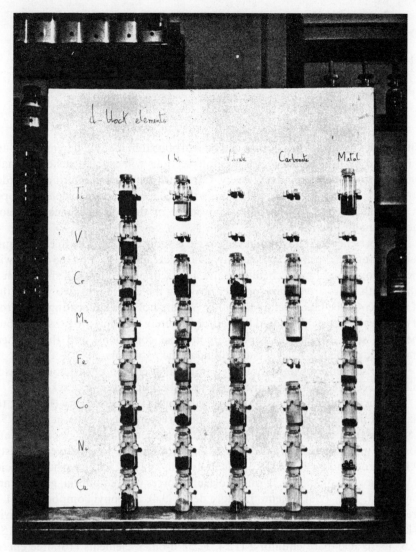

Figure A22
A chemical specimen display board.

Appendix 4
Possible alternative pathways through the course

The subject matter of this course can be rearranged in many different ways, and it is hoped that the division of topics into sections will enable teachers to make such rearrangements relatively easily. Care must be taken in altering the order, however, for many ideas in each topic as written depend upon concepts developed in preceding topics. A thorough knowledge of the course is needed before tackling major alterations.

Experience has shown that the course is sufficiently flexible to enable quite major changes to be made successfully. Examples include starting the course with the first topic on organic chemistry and, in a school with a small sixth form in which both years are taught together, dividing the course into two years either of which could be taken before the other. We are not here advocating that either of these rearrangements should be made, but merely that they are possible should the need arise. Much will depend upon local conditions, and the teachers' knowledge of the pre-A-level background of their students.

The sequence of topics followed in the first edition of this course was as follows (second edition topic and section numbers are given in brackets).

General introduction, amount of substance (1.1); periodicity (1.2, 5.4); the masses of molecules and atoms (3); atomic structure (4); the halogens (5.2, 5.3, 5.5, 5.6); oxidation number (5.1); the s-block elements (2); energy changes and bonding (6); structure (7); bonding (8); introduction to organic chemistry (9.1); hydrocarbons (9.2, 9.4, 9.5); alcohols, phenol (11.1, 11.2); carboxylic acids (13.1); halogenoalkanes (9.3); intermolecular forces (10); equilibria, gaseous and ionic (12.1, 12.2, 12.3); alkenes in more detail (9.4); carbonyl compounds (11.3); arenes in more detail (9.5); amines (13.3); amino acids (13.4); optical activity (11.4); a problem in synthesis (17.5); reaction rates (14); redox equilibria (15.2, 15.3, 15.4); acid–base equilibria (12.4, 12.5); d-block elements (16); free energy and related topics (15.1, 15.5); proteins (13.4); polymers (17.1); nitrogen and sulphur (18).

If this sequence is preferred, attention should be given to the diagram on page xv of *Teachers' guide I*, which shows the position of the various sections on the Second Law of Thermodynamics, no treatment of which appears in the first edition, so that they can be fitted in at suitable points.

A further factor which should be borne in mind when deciding upon a different sequence is the choice of Special Study, and the position in the sequence that it is proposed to teach it. Foundations must be laid for each of the Studies; for example, amino acids, proteins, and carbohydrates should be studied before the *Biochemistry* Special Study is attempted.

Reference sources and bibliography

The books which are listed here are intended as a general guide to some sources of further information which teachers may from time to time find it helpful to consult. There are also some books to which the teacher is referred for further reading on particular aspects of individual topics; these are listed both here and in the topics concerned. The works listed here are arranged under a series of headings, beginning with general sources of information for the course as a whole, and continuing with individual subject headings arranged alphabetically. The list is not claimed to be exhaustive; it is merely a collection of some sources which have been found to be helpful.

General chemistry

The Royal Society of Chemistry publish a number of monographs for teachers, and special reports of interest to teachers. Details of titles currently available can be obtained from the Marketing Department, The Royal Society of Chemistry, Burlington House, London W1V 0BN.

The Open University course books on a number of chemical topics are of value to teachers at this level. Details can be obtained from Open University Educational Enterprises Ltd., 12 Cofferidge Close, Stony Stratford, Milton Keynes MK11 1BY.

Television programmes accompanying the courses are also of value; details of transmission times of programmes of chemical interest are published each year in the January issue of *Education in Chemistry*. A licence is required to videotape these programmes for class use; for details apply to Guild Learning Licensing Department, Guild House, Peterborough PE2 9PZ.

The Inner London Project for Advanced Chemistry (ILPAC) is a highly detailed course of study for students working towards A-level, and provides valuable resource material for teachers. It is presented to students in twenty guides covering the two years' work, and the guides are accompanied by teachers' and technicians' notes, ILPAC is published by John Murray, 50 Albemarle Street, London W1X 4BD.

Applications of chemistry

BRITISH PETROLEUM. *Our industry: petroleum.* 1977.
SELINGER, B. *Chemistry in the marketplace.* John Murray, 1979.
WADDAMS, A.L. *Chemicals from petroleum.* John Murray, 1978.

Chemical nomenclature and units

ASSOCIATION FOR SCIENCE EDUCATION. *SI units, signs, symbols, and abbreviations.* 1981.
ASSOCIATION FOR SCIENCE EDUCATION. *Chemical nomenclature, symbols, and terminology.* 1984.
CAHN, R.S. *Introduction to chemical nomenclature.* Butterworth, 5th edn, 1979.

Chemical energetics and equilibrium

ATKINS, P.W. *Physical chemistry.* Oxford University Press, 2nd edn, 1982.
BENT, H.A. *The second law.* Oxford University Press, New York, 1965.
REIF, F. Berkeley Physics Course. Volume 5 *Statistical physics.* McGraw Hill, 1964.
REVISED NUFFIELD ADVANCED PHYSICS. *Students' guide 2* and *Teachers' guide 2* Unit K, 'Energy and entropy'. Longman, 1985.

Chemical kinetics

DAWSON, B.E., MASON, C.L., and MASON, P. *Reaction kinetics: a resource for teachers.* Royal Society of Chemistry, 1981.

Data books

AYLWARD, G.H. and FINDLAY, T.J.V. *SI chemical data.* John Wiley Australasia Pty Ltd, 1976.
REVISED NUFFIELD ADVANCED SCIENCE. *Book of data.* Longman, 1984.
WEAST, R.C. (ed) *Handbook of chemistry and physics.* Chemical Rubber Company Press, Cleveland, Ohio; Blackwell, Oxford.

History of chemistry

KNIGHT, D.M. (ed.) *Classical scientific papers: chemistry (second series).* Bell and Hyman, 1970.

Inorganic chemistry

COTTON, F.A. and WILKINSON, G. *Advanced inorganic chemistry.* Wiley Interscience, 4th edn, 1980.
NICHOLLS, D. *Complexes and first-row transition elements.* Macmillan, 1975.
PHILIPS, C.S. and WILLIAMS, R.J. *Inorganic chemistry* (2 volumes). Oxford University Press, 1965.

Journals

School science review, published quarterly by the Association for Science Education.
Education in chemistry, published bi-monthly by the Royal Society of Chemistry.
Journal of chemical education, published monthly by the Division of Chemical Education of the American Chemical Society.
Scientific American, published monthly. Selected articles can be obtained as *Scientific American offprints*.

Learned societies

The Association for Science Education, College Lane, Hatfield, Hertfordshire AL10 9AA.
The Royal Society of Chemistry, Burlington House, London W1V 0BN.
   The Royal Society of Chemistry operates an 'affiliation scheme' for schools, under which participating chemistry departments obtain *Education in chemistry*, and other periodicals and information, at favourable rates. For details of this scheme, apply to the Education Officer.

Organic chemistry

FIESER, L.F. and WILLIAMSON, K.L. *Organic experiments*. Heath, 4th edn, 1979.
FINAR, I.L. *Organic chemistry* (2 volumes). Longman, 1975.
MANN, F.G. and SAUNDERS, B.C. *Practical organic chemistry*. Longman, 4th edn, 1979.
NORMAN, R.O.C., TOMLINSON, M.J., and WADDINGTON, D.J. *Mechanisms in organic chemistry: case studies*. Bell and Hyman, 1978.
OPENSHAW, H.T. *A laboratory manual of qualitative organic analysis*. Cambridge University Press, 3rd edn, 1976.
ROBERTS, J.D. and CASERIO, M.C. *Basic principles of organic chemistry*. Benjamin, 1977.
SYKES, P. *A guidebook to mechanism in organic chemistry*. Longman, 5th edn, 1981.
SYKES, P. *The search for organic reaction pathways*. Longman, 1972.
Vogel's textbook of practical organic chemistry. Longman, 4th edn, 1978.

Safety

ASSOCIATION FOR SCIENCE EDUCATION. *Safeguards in the school laboratory*. 8th edn, 1981.
ASSOCIATION FOR SCIENCE EDUCATION. *Topics in safety*. 1982.
BRETHERICK, L. *Hazards in the chemical laboratory*. Royal Society of Chemistry, 3rd edn, 1981.
CLEAPSE and SSERCE *CLEAPSE hazards*. 1981. Available from the Association for Science Education.

DEPARTMENT OF EDUCATION AND SCIENCE. *Safety in science laboratories.* DES Safety series no. 2. HMSO, 2nd edn, 1983.

EVERETT, K. and JENKINS, E.W. *A safety handbook for science teachers.* John Murray, 3rd edn, 1980.

SCOTTISH SCHOOLS SCIENCE EQUIPMENT RESEARCH CENTRE. *Hazardous chemicals – a manual for schools and colleges.* Oliver and Boyd, 1979.

Structure and bonding

'The architecture of matter: A-level series.' Filmstrips or slide sets (about 100 frames) on structure problems. Available from Dr M. B. Ormerod, Scientific Educational Aids, 104 Hercies Road, Uxbridge, Middlesex UB10 9ND.

DYKE, S.F., FLOYD, A.J., SAINSBURY, M., and THEOBALD, R.S. *Organic spectroscopy.* Longman, 1978.

GILBERT, B.C. *Investigation of molecular structure.* Bell and Hyman, 1984.

WELLS, A.F. *Structural inorganic chemistry.* Oxford University Press, 4th edn, 1975.

Schools which took part in the trials

The Nuffield–Chelsea Curriculum Trust puts on record its warmest thanks to the following schools, and to their teachers whose advice, assistance, and support in the development of the course was greatly valued.

Schools which took part in the trials of the original course

From September, 1966
Blyth Grammar School
Braintree County High School
Deeside Senior High School, Flints.
Edinburgh Academy
Huddersfield New College
King's College School, Wimbledon
Methodist College, Belfast
Nottingham High School for Girls
Pontefract Girls' High School, Yorks
Prendergast Grammar School,
    Catford, London SE6
Royal Grammar School, High
    Wycombe
Westminster School

Additional schools from September, 1967
Bradford Girls' Grammar School
Clifton College, Bristol
Grosvenor High School, Belfast
King Edward VI Grammar School,
    Aston, Birmingham
Maidstone Grammar School
Malvern College
Newport High School
Oundle School
Rossall School, Fleetwood, Lancs.

Schools which took part in the trials of the revised topics

Archbishop Holgate's School, York
Bath High School G.P.D.S.T.
Bishop Walsh School, Sutton Coldfield
Castle School, Thornbury
Charterhouse
Christ's Hospital
Durrants School, Croxley Green
Harvey Grammar School, Folkestone
Helenswood School, Hastings
Ipswich School
Rickmansworth School
W.R. Tuson College, Preston.

Index

References to specific salts are indexed under the name of the appropriate cation; references to substituted organic compounds are indexed under the name of the parent compound.

a

abbreviations, 193
Acid Blue, 40, 159
acid–base equilibria, 16–24
acids, K_a of weak, 26–7
   strengths, 17–19
acrylic polymers, 156, 157
activation energy, 79, 80–81
activity coefficients, 127–8
alkanes, halogeno-, hydrolysis mechanisms, 69–71
aluminium compounds, 179–80
amides, *see under* carboxylic acids, derivatives
amines, 48–50
   diazotization, 158–9
amino acids, 50–54
ammonia
   ammonia–ethanoic acid titration, 22
   ammonia–hydrochloric acid titration, 21–2
   ammonia–nitrogen–hydrogen equilibrium, 13, 29–30
   distribution between solvents, 4–5
   model, 214
   reactions, 186
ångstrom, 193
'anti-clockwise rule', 119
apparatus, 194–211
argon, determination of ionization energy, 196–9
aspirin, 160, 161
atmosphere (unit), 193
autocatalysis, 82, 85

b

ball-and-spring models, 212–17
bases, *see* acid–base equilibria
benzenediazonium chloride, decomposition, 80
benzocaine, synthesis, 167
benzoic acid, esterification, 44
   identification, 163–4
   K_a, to measure, 26–7
   reduction, 45
benzoic acid, 2-hydroxy-, syntheses using, 161
benzoic acid, 4-methoxy-, model, 215
benzoyl chloride, precautions in use, 47
beryllium chloride, model, 214
biuret test for peptide groups, 52
body-centred cubic structure, model, 218
boron compounds, 179–80
boron trifluoride, model, 214
Brady's reagent, *see* hydrazine, 2,4-dinitrophenyl-
bromine–hydrogen–hydrogen bromide equilibrium, 12
bromine reagent, preparation, 168
bromophenol blue, 27
   K_a, to measure, 25–6
buffer solutions, 24–5
butane, 1-bromo-, hydrolysis mechanisms, 70
butanone, identification, 163–4, 165
butylamine, reactions, 48–50

c

caesium chloride, model, 216, 217, 218
calcium carbonate
   calcium carbonate–calcium oxide–carbon dioxide equilibrium, 15
   calcium carbonate–hydrochloric acid reaction, kinetic study, 66–8
   decomposition, 125
calcium fluoride, *see* fluorite
calcium oxide–carbon dioxide–calcium carbonate equilibrium, 15
calorimeter, electrical compensation, 200–201
carbamide, enzymic hydrolysis, 55–6
carbon
   carbon–hydrogen–methane equilibrium, 30–31
   reactions with water, 125
carbon dioxide
   carbon dioxide–nitrogen monoxide reaction, 13
   carbon dioxide–calcium oxide–calcium carbonate equilibrium, 15
   model, 215
carbon monoxide
   carbon monoxide–hydrogen–methanol equilibrium, 12
   carbon monoxide–nitrogen dioxide reaction, 13
carbonates, reaction with carboxylic acids, 44
carbonyl compounds, identification, 162–3
carboxylic acids, 43–5
   derivatives, 45–7, 161
catalysis, 81–5
catalysts, transition elements as, 150–51
   see also enzymes
cell diagrams, 99
cellulose ethanoate (acetate), 157, 159
chemical kinetics, *see* reaction rates
chemical specimens display board, 221–2, 223
chirality, 51, 52

chlorides, p-block elements, 178–80
chlorine, model, 213
cholesteryl benzoate, preparation, 47
chromium(II) ethanoate, preparation, 147–9
collision theory of reaction kinetics, 73–81
colorimeter, 205–8
  reaction monitoring using, 83–4
combustion analysis, 162
complex ions, see metal complexes
concentrations, effect on electrode potentials, 105–13
  effect on reaction rates, 65–71
copper, reactions: with hydrogen chloride, 94; with nitric acid, 185; with silver(I) ion, 95–6
copper complexes, 145–6, 186
copper(II) ion, reactions: with ammonia solution, 186; with zinc; 95–6, see also Daniell cell
copper(I) thiocarbamide chloride, preparation, 147–9
cotton, 157, 159
course pathways, 224–5
crossed polaroid assembly, 202–3
crystal structures, models, 216–18

d

Daniell cell, 98–9, 130–31
derivatization, 162–3
diamond, model, 216
diazonium compounds, 80, 159
diazotization, 48, 158–9
dinitrogen tetraoxide–nitrogen dioxide equilibrium, 12
Direct Red, 23, 159
Disperse Yellow, 3, 159
display boards, 220–23
disproportionation, sulphur compounds, 188
dissociation constants, to measure, 23–4
  pK_a, 28
drugs, 160–62
dyes, 158–60

e

electrical compensation calorimeter, 200–201
electrode potentials, effect of concentration on, 105–13
  to measure, 115–16
  see also standard electrode potentials
electrodes, contribution to e.m.f., 99
  hydrogen, 99–102, 109, 209–11
  reference, 99, 102
  silver, 106–8
electromotive force of cell, 99
  to calculate, 118
  to measure, 100–105; potentiometer for, 208–9
  uses, 122
electron transfer, 96–7

Ellingham diagrams, 151
end-point, 20
energy levels, display boards, 220–22
  transition elements, 139–40
energy of activation, 79, 80–81
entropy 'balance sheet', 29, 30–31
entropy changes, 93–4, 124–32
  in reactions of complexes, 150
  in voltaic cells, 120–22
  zero, at equilibrium, 29–31
enzymes, 55–6
equilibria, 1–41
  acid–base, 16–24
  effect of pressure and temperature, 10–14
  heterogeneous, 14–16
  relative concentrations, 4–9
  zero total entropy change at, 29–31
  see also redox reactions
equilibrium constant, 6, 9–10, 14, 122
  in terms of partial pressures, 10–12
  to calculate, 124
  to find, using Nernst equations, 119
Equilibrium Law, 2–10
esters, preparation, 161
  see also under carboxylic acids, derivatives
ethane
  ethane–ethene–hydrogen equilibrium, 13
  model, 214
ethane, 1,1,1-trichloro-, precautions in use, 4–5
ethanedioic acid, dehydration of hydrated, 168
ethanedioic acid–manganate(VII) reaction, kinetic study, 82–5
ethanoic acid, esterification, 4, 7–9
  ethanoic acid–ammonia titration, 22
  ethanoic acid–sodium hydroxide titration, 22
  reactions, 44
ethanoic acid, amino-, pH of solution, 18–19
ethanol, reaction with ethanoic acid, 4, 7–9
ethanoyl chloride, precautions in use, 46–7, 49–50
ethanoylation, 160–61
ethene
  ethene–hydrogen–ethane equilibrium, 13
  model, 214
  reaction with hydrogen, 94
ethyl ethanoate, hydrolysis, 4, 7–9
ethyne, model, 215
'expansion of the octet', 178, 187
extensive properties, 3

f

face-centred cubic structure, model, 218
Fehling's solution, 163
'ferrous sulphate' tablets, analysis, 143–4
fibres, choice of, 157
  see also textiles
fluorite, model, 216, 217

free energy chance, 83–4, 120
  see also standard free energy change

g

gas constant (R), 11
gas syringes, 194
Gibbs free energy change, see free energy change
glycine, see ethanoic acid, amino-
graphite, model, 215

h

half-equations, 97
heterogeneous equilibria, 14–16
hexagonal close-packed cubic structure, model, 218
hydrazine, 2,4-dinitrophenyl-, 162–3
hydrochloric acid, see hydrogen chloride
hydrogen
  hydrogen–bromine–hydrogen bromide equilibrium, 12
  hydrogen–carbon–methane equiligrium, 30–31
  hydrogen–carbon monoxide–methanol equilibrium, 12
  hydrogen–ethene–ethane equilibrium, 13
  hydrogen–iodine–hydrogen iodide equilibrium, 4, 5–7, 13
  hydrogen–nitrogen–ammonia equilibrium, 13, 29–30
  reactions: with ethene, 94; with nitrogen, 125; with silicon, feasibility, 94
hydrogen bromide–hydrogen–bromine equilibrium, 12
hydrogen chloride
  hydrochloric acid–ammonia titration, 21–2
  hydrochloric acid–sodium hydroxide titration, 21, 23–4
  model, 214
  reactions: with calcium carbonate, kinetic study, 66–8; with copper, 94; with sodium thiosulphate, kinetic study, 72–3
hydrogen cyanide, model, 215
hydrogen electrode, 99–102, 109, 209–11
hydrogen fluoride, model, 214
hydrogen iodide, decomposition, 80
  hydrogen iodide–hydrogen–iodine equilibrium, 4, 5–7, 13
hydrogen sulphide, electronic structure, 187
  model, 214
  precautions in use, 188
hypodermic syringes, 194

i

ice, lead in, 183
ideal gas equation, 10–11
indicators, for acid–base titrations, 20–21
  K_a, 25–6
indigo, synthesis, 167

infra-red spectroscopy, 193
intensive properties, 3
iodide ion
  iodide–iodine equilibrium, 115–16
  iodide–iron(III) reaction, 114–15
iodine
  iodine–hydrogen–hydrogen iodide equilibrium, 4, 5–7, 13
  iodine–iodide equilibrium, 115–16
  iodine–propanone reaction, kinetic study, 68–9, 85
ion/ion systems, 113–19
ionization energy, determination, 196–9
iron, determination in 'ferrous sulphate' tablets, 143–4
  iron(III) ion–iodide reaction, 114–15
  iron(II)–iron(III) equilibrium, 115–18
  redox reactions, 141–2
iron complex, 147

j

joule, 193

k

kelvin, 193

l

law of partial pressures, 10
lead, pollution by, 183–4
lead/copper cell, 121
lead(IV) oxide, 181–3
lead(II) sulphate, reaction with ammonia solutions, 186
Le Chatelier's principle, 13–14
length, units, 192–3
liquid crystals, preparation, 47
lithium tetrahydridoaluminate(III) (lithium aluminium hydride) reductions, 45
litmus, 27
litre, use of term, 193

m

magnesium sulphate, reaction with ammonia solutions, 186
manganate(VII) ion, reactions: with ethanedioic acid, kinetic study, 82–5; with nitrous acid, 185; with sulphite ion, 188
manganese(II) ion, catalysis by, 82–3, 85
mass spectrometer, demonstration of principle, 199–200
mass spectrometry, 162, 164–5
melting point, determination, 50
metal complexes, 145–50, 186
metal/metal ion systems, 95–105
  see also electrode potentials
methane
  methane–carbon–hydrogen equilibrium, 30–31
  model, 214

methanoic acid, pH of solution, 19
methanol–carbon monoxide–hydrogen equilibrium, 12
methanol, phenyl-, preparation, 45
methyl benzoate, preparation, 44
methyl orange, 27
micro-electronics, 183
models, 212–19
molecular formula, determination, 162
molybdenum, oxidation numbers, 143

n
nanometre, 193
naphthalen-2-ol, 158–9
Nernst equation, 110, 117, 119, 129–31
newton, 193
nickel complex, 147
ninhydrin test for amino acids, 52
nitric acid, precautions in use, 181
  reactions, 185
nitrogen
  nitrogen–hydrogen reaction, 125
  nitrogen–hydrogen–ammonia equilibrium, 13, 29–30
  nitrogen–oxygen–nitrogen monoxide equilibrium, 13
nitrogen compounds, 184–6
nitrogen dioxide
  nitrogen dioxide–carbon monoxide reaction, 13
  nitrogen dioxide–dinitrogen tetraoxide equilibrium, 12
nitrogen monoxide
  nitrogen monoxide–carbon dioxide equilibrium, 13
  nitrogen monoxide–nitrogen–oxygen equilibrium, 13
nitrous acid, reactions, 158–9, 185–6
nomenclature, 192
nylon, 157, 159

o
organic compounds, identification, 162–5
overhead projection originals, 219–20
oxidation by electron transfer, 96–7
oxidation numbers, 187
  display board, 222
  vairable, 141–5
oxides, p-block elements, 178–9
oxygen
  oxygen–nitrogen–nitrogen monoxide equilibrium, 13
  oxygen–sulphur dioxide–sulphur trioxide equilibrium, 12, 13

p
paper chromatography of amino acids, 53
PEEL models, 219

peptides, 51
Perspex, 156–7
persulphate ion, electronic structure, 187
pH, changes during titrations, 19–24
  of strong and weak acids, 17–19
pH meter, 17, 21, 112–13
phenolic resin, 156–7
phenolphthalein, 27
phenylamine, precautions in use, 49
  reactions, 48–50
phosphine, model, 214
phosphoric(v) acid, catalyst, 161
phosphorus, model, 213
phosphorus(v) oxide, model, 213
phosphorus pentachloride, model, 213
polarimeter, 203–5
polaroids, crossed, assembly, 202–3
pollution, lead, 183–4
polyesters, 157–159
  resin, 156–7
polymers, 155–8
poly(methyl 2-methylpropenoate), 156–7
poly(propenamide), 156–7
potassium dichromate(VI), 148
potassium iodide, reactions: with nitric acid, 185; with nitrous acid, 185; with sulphite ion, 188
potentiometer, 208–9
prefixes, unit names, 192–3
pressure, units, 193
propane, 2-bromo-2-methyl-, hydrolysis mechanism, 70–71
propan-2-ol, identification, 163–4, 165
propanone–iodine reaction, kinetic study, 68–9, 85
proteins, 50–54
pyridine, precautions in use, 47

r
radio valves, for ionization energy determination, 196–9
rate equations, 65–6
rayon, 157
reaction rates, 63–4
  effect of concentration, 65–71
  effect of temperature, 71–81
redox reactions, equilibria, 113–19
  of iron, 141–2
  of vanadium, 142–3
  to predict likelihood, 118–19, 122
  see also metal/metal ion systems
reduction by electron transfer, 96–7
reference electrode, 99, 102
reference sources, 226–9
resins, 156–7
reversible reactions, 3

s

scandium, electronic structure, 139
silane, model, 214
silicon, reaction with hydrogen, feasibility, 94
silicon dioxide, in micro-electronics, 183
silicon tetrachloride, model, 213
silk, 157
silver, determination from e.m.f. measurement, 111–12
silver/copper cell, 121
silver electrode, effect of silver ion concentration, 106–8
silver(I) ion–copper reaction, 95–6
sodium chloride, model, 216, 217, 218
sodium dichromate(VI) reagent, preparation, 168
sodium ethanoate, dehydration of hydrated, 168
sodium hydroxide
  sodium hydroxide–ethanoic acid titration, 22
  sodium hydroxide–hydrochloric acid titration, 21, 23–4
sodium thiosulphate, disproportionation, 188
  reaction with hydrochloric acid, kinetic study, 72–3
solubility product, 15, 16, 111
space-filling models, 218–19
spectroscope, direct-vision, 194–5
standard conditions, 94, 123, 124
standard electrode potentials, 102, 109
  uses, 118–19
standard free energy change, 29, 122
standard free energy of formation, 123–32
steroid, synthesis, 167
sulphate ion, electronic structure, 187
sulphide ion, reactions, 188
sulphite ion, electronic structure, 187
  reactions, 188
sulphur, model, 213
sulphur compounds, 186–9
sulphur dioxide
  precautions in use, 188
  reaction with nitric acid, 185
  sulphur dioxide–oxygen–sulphur trioxide equilibrium, 12, 13
synthesis, problems in, 166–8

syringes, 194

t

teaching sequence, *see* course pathways
temperature, units, 193
textiles, dyeing of, 159
tin(IV) oxide, 181–2
titrations, acid–base, 19–24
transition elements, 137–53
  definition, 138, 139
transparencies, 220

u

units, 192–3
urea, 55–6
  thiourea, 148, 149
urease, 55–6

v

valves, *see* radio valves
vanadium, redox reactions, 142–3
voltaic cells, 97–9, 114–16
  see also electromotive force of cell
volume, units, 193

w

waste liquids, disposal, 4–5
water, ionization constant, 16, 17
  model, 213
  reaction with carbon, 125
wavenumber, 193
wool, 157, 159

x

xenon, determination of ionization energy, 196–9

z

Zartman experiment, 77
zinc, reaction with copper(II) ion, 95–6
  see also Daniell cell
zinc blende, model, 216, 217
zinc sulphate, reaction with ammonia solution, 186
zinc sulphide, *see* zinc blende